**Redox-Mediated Signal Transduction**

# METHODS IN MOLECULAR BIOLOGY™

## *John M. Walker,* SERIES EDITOR

METHODS IN MOLECULAR BIOLOGY™

# Redox-Mediated Signal Transduction

## Methods and Protocols

Edited by

## John T. Hancock, PhD

*Centre for Research in Plant Science, University of the West of England,*
*Bristol, UK*

 Humana Press

*Editor*
John T. Hancock, Ph.D.
Centre for Research in Plant Science
University of the West of England
Bristol
UK


*Series Editor*
John M. Walker
School of Life Sciences
University of Hertfordshire
Hatfield, Hertfordshire AL10 9AB
UK

ISBN: 978-1-58829-842-3          e-ISBN: 978-1-59745-129-1
ISSN: 1064-3745                  e-ISSN: 1940-6029
DOI: 10.1007/978-1-59745-129-1

Library of Congress Control Number: 2008936257

9 8 7 6 5 4 3 2 1

springer.com

# Preface

Redox biology has been studied for a long time, with some of the most well known redox mechanisms being amongst the most important cellular functions. Historically, these include those that revolve around energetics, including the production of ATP. However, more recently, the role of redox in the control of cellular activity has become more prominent, with the realisation that cells not only produce compounds that can affect the intracellular and extracellular redox, but that the alterations of the redox of cellular compartments can control enzyme functions. Therefore it is timely to produce a book which brings together some of the methods and techniques that can be used to study this important aspect of biology.

I am of course indebted to all those that have taken time out of their busy lives to contribute chapters to this book, and without them it simply would not have been possible, so a big thank you to all. The book starts with a overview, and then moves on to methods for measuring compounds that affect redox and the redox state of the cells. Clearly this is not an all encompassing list of methods, but a selection of ones currently being used. The use of GFP, and its derivatives, is still being developed and chapters on this topic are more in the style of reviews, hopefully giving the reader an idea of what is possible with such technology. The book then moves on to the methods to study the impact of changing redox on proteins, including how to identify proteins that can be affected in this way, and methods to study the exact molecular changes that might underlie the mechanisms of action of altering redox. Chapters also look at global changes in cells.

The authors who kindly contributed to this book were drawn from researchers in both the animal and the plant fields. Many of the techniques used can be adapted for whichever organisms, tissues or cell types are being studied, but often plant specific or animal-specific points need to be considered, and where possible authors have highlighted this.

This book is part of a large series, and there is no doubt that it will overlap with others in the series, but these chapters have all been written by authors who have a wide experience on working in redox biology, and it is hoped that it will be useful not only to those who wish to enter this exciting area of research, but also to those that wish to expand their repertoire of techniques so that a more full understanding of the impact of redox in the control of cellular function can be reached.

Finally I would like to thank the series editor, John Walker, for asking me to edit this book, and my family for their continual support in such projects.

*John T. Hancock*

# Contents

# Contributors

JEFFREY S. ARMSTRONG • *Department of Biochemistry, National University of Singapore, Singapore*

DAVID BARFORD • *Section of Structural Biology, Institute of Cancer Research, Chester Beatty Laboratories, London, UK*

VSEVOLOD V. BELOUSOV • *Shemyakin-Ovchinnikov Institute of Bioorganic Chemistry, Russian Academy of Sciences, Moscow, Russia*

EKATERINA A. BOGDANOVA • *Laboratory of Molecular Technologies for Biology and Medicine, Shemyakin-Ovchinnikov Institute of Bioorganic Chemistry, Russian Academy of Sciences, Moscow, Russia*

D. ALLAN BUTTERFIELD • *Department of Chemistry, Center of Membrane Sciences and Sanders-Brown Center on Aging, University of Kentucky, Lexington, KY, USA*

MARK B. CANNON • *Department of Biochemistry and Physical Sciences, BYU-Hawaii, Laie, HI, USA*

STEVEN COLES • *Centre for Research in Biomedicine, University of the West of England, Bristol, UK*

MYRA E. CONWAY • *Centre for Research in Biomedicine, University of the West of England, Bristol, UK*

ROBERT C. CUMMING • *Department of Biology, University of Western Ontario, London, Ontario, Canada*

LAURA DE GARA • *Dipartimento di Biologia e Patologia Vegetale, Bari, Italy*

RADHIKA DESIKAN • *Division of Biology, Imperial College London, London, UK*

YUKTEE DOGRA • *Institute of Biomedical and Clinical Science, Peninsula Medical School, Exeter, UK*

JÖRG DURNER • *Institute of Biochemical Plant Pathology, GSF - National Research Center for Environment and Health, Munich/Neuherberg, Germany*

CHRISTINE H. FOYER • *School of Agriculture, Food and Rural Development, Newcastle University, Newcastle upon Tyne, UK*

JOHN T. HANCOCK • *Centre for Research in Plant Science, University of the West of England, Bristol, UK*

QUANZHEN HUANG • *Department of Chemistry, Center of Membrane Sciences and Sanders-Brown Center on Aging, University of Kentucky, Lexington, KY, USA*

SUSAN M. HUTSON • *Wake Forest University Health Sciences, Winston-Salem, NC*

THOMAS E. INGRAM • *Department of Cardiology, Wales Heart Research Institute, Cardiff University Medical School, Cardiff, UK*

PHILIP E. JAMES • *Department of Cardiology, Wales Heart Research Institute, Cardiff University Medical School, Cardiff, UK*

AFSHIN KHALATBARI • *Department of Cardiology, Wales Heart Research Institute, Cardiff University Medical School, Cardiff, UK*

TRACY LAWSON • *Department of Biological Sciences, University of Essex, Colchester, Essex, UK*

NATACHA LE MOAN • *CEA, DSV, IBITECS, SBIGEM, Laboratoire Stress Oxydants et Cancer, Gif-sur-Yvette, France*

CHRISTIAN LINDERMAYR • *Institute of Biochemical Plant Pathology, GSF - National Research Center for Environment and Health, Munich/Neuherberg, Germany*

VITTORIA LOCATO • *Dipartimento di Biologia e Patologia Vegetale, Bari, Italy*

SERGEI LUKYANOV • *Laboratory of Molecular Technologies for Biology and Medicine, Shemyakin-Ovchinnikov Institute of Bioorganic Chemistry, Russian Academy of Sciences, Moscow, Russia*

KRISTINE MANN • *Department of Biological Sciences, University of Alaska Anchorage, Anchorage, AK, USA*

KSENIYA N. MARKVICHEVA • *Laboratory of Molecular Technologies for Biology and Medicine, Shemyakin-Ovchinnikov Institute of Bioorganic Chemistry, Russian Academy of Sciences, Moscow, Russia*

PHILIP M. MULLINEAUX • *Department of Biological Sciences, University of Essex, Colchester, Essex, UK*

STEVEN J. NEILL • *Centre for Research in Plant Science, University of the West of England, Bristol, UK*

SHELLEY F. NEWMAN • *Department of Chemistry, Center of Membrane Sciences and Sanders-Brown Center on Aging, University of Kentucky, Lexington, KY, USA*

TILL K. PELLNY • *Crop Performance and Improvement Division, Rothamsted Research, Harpenden, Herts, UK*

ANDREW G. PINDER • *Department of Cardiology, Wales Heart Research Institute, Cardiff University Medical School, Cardiff, UK*

LESLIE B. POOLE • *Wake Forest University Health Sciences, Winston-Salem, NC*

S. JAMES REMINGTON • *Institute of Molecular Biology and Department of Physics, University of Oregon, Eugene, OR*

STEPHEN C. ROGERS • *Department of Cardiology, Wales Heart Research Institute, Cardiff University Medical School, Cardiff, UK*

ANNETTE SALMEEN • *Department of Chemical and Systems Biology, Stanford University Medical Center, Stanford, CA*

SIMONE SELL • *Institute of Biochemical Plant Pathology, GSF - National Research Center for Environment and Health, Munich/Neuherberg, Germany*

JENNA SLINN • *Faculty of Health and Life Sciences, University of the West of England, Bristol, UK*

DMITRY B. STAROVEROV • *Evrogen, JSC, Moscow, Russia*

RUKHSANA SULTANA • *Department of Chemistry, Center of Membrane Sciences and Sanders-Brown Center on Aging, University of Kentucky, Lexington, KY, USA*

FRÉDÉRIQUE TACNET • *CEA, DSV, IBITECS, SBIGEM, Laboratoire Stress Oxydants et Cancer, Gif-sur-Yvette, France*

MICHEL B. TOLEDANO • *CEA, DSV, IBITECS, SBIGEM, Laboratoire Stress Oxydants et Cancer, Gif-sur-Yvette, France*

MATT WHITEMAN • *Institute of Biomedical and Clinical Science, Peninsula Medical School, Exeter, UK*

PAUL G. WINYARD • *Institute of Biomedical and Clinical Science, Peninsula Medical School, Exeter, UK*

# Chapter 1

# The Role of Redox in Signal Transduction

## John T. Hancock

## Abstract

Functioning and efficient cell signaling is vital for the survival of cells. Over the course of many years, various components have been identified and recognized as crucial for the transduction of signals in cells. Many of the mechanisms allow for a relatively rapid switching of signals, on or off, with common examples being the G proteins and protein phosphorylation. However, recently it has become apparent that other modifications of amino acids are also important, including reactions with nitric oxide, for example, S-nitrosylation, and of particular relevance here, oxidation of cysteine residues. Such oxidation will be dependent on the redox status of the intracellular environment in which that protein resides, and this will in turn be dictated by the presence of pro-oxidants and antioxidants. Here, the chemistry of redox modification of amino acids is introduced, and a general overview of the role of redox in mediating signal transduction is given.

**Keywords:** Cysteine modification, hydrogen peroxide, nitric oxide, redox, S nitrosylation, signal transduction, thiol.

## 1. Introduction

Redox chemistry often is frightening for those that are not passionate about the topic. Students, and indeed researchers, often shy away from lectures or seminars on redox, or they simply cannot comprehend the relevance of these principles and formulae for what they are hoping to become in the future. However, when considered at the basic level, redox simply means that there is reduction and oxidation of compounds taking place, that is, there is simply the movement of electrons. Many of the reactions that allow an organism to survive are based on redox. Mitochondria

John T. Hancock (ed.), *Methods in Molecular Biology, Redox-Mediated Signal Transduction, vol. 476*
© 2008 Humana Press, a part of Springer Science + Business Media, Totowa, NJ
DOI: 10.1007/978-1-59745-129-1_1

make ATP, our ubiquitous source of useful energy, by harnessing the power of redox. Chloroplasts are driven by redox.

Historically, this is where the world of redox has been residing, in those laboratories around the world that are interested in energy. How can we take the sun's energy and make an organism thrive and grow, and how can that healthy lunch be transformed into useful energy to drive our muscles? However, during the years redox has been viewed to be more than this, and it was found that redox was being used to rid bodies of xenobiotic materials, via such systems as the P450 complex. Pathogens could be warded off by redox enzymes such as the NADPH oxidase complex.

However, the mid-1980s marked a new era beginning for the redox world. It was found that a molecule previously referred to as endothelial-derived relaxing factor (EDRF) was in fact a gaseous free radical molecule and one that would therefore partake in redox chemistry. That molecule was nitric oxide (NO). Once researchers realized that NO was taking part in signaling, the door was opened to ask the question about the involvement of redox in general. Many other groups, who had an interest in the roles of reactive oxygen species (ROS, sometimes referred to as active oxygen species) started to look to see whether the manipulation of ROS levels in and around cells had effects on the signaling that ensued, and many discovered that ROS could be considered as key signaling components in many pathways (1–4).

Here, the notation for nitric oxide of NO is used. However, it should be noted that the radical form of NO should be written as NO$^{•}$. Furthermore, with gain or loss of electrons NO can also be NO$^{-}$ and NO$^{+}$. It cannot be assumed that NO donors release NO$^{•}$; some release NO$^{+}$, for example, so when NO donors are used, the exact chemistry needs to be checked and understood before data are interpreted.

## 2. Cell Signaling

Cell signaling is one of the most important aspects of modern biological sciences. It is clear that for a cell to survive, it has to be able to respond to its environment and change its activity accordingly. Many mechanisms exist to perceive the presence of extracellular factors, for example, the arrival of hormones or cytokines, or a change in environmental conditions such as temperature, and yet further mechanisms are in place to transmit a signal deep into the cell. The resultant changes in activity may be in a range of places in the cell, including modulation of metabolic activity in the cytoplasm, or the up- or downregulation of gene expression in the nucleus (for example *see* **ref.** *5*).

No matter what initiates the response, or in fact where the response is finally observed, one of the key mechanisms used by cell signaling pathways is the rapid and reversible turning on or off of protein function. Some cell signaling components have indeed been referred to as molecular switches, such as the G proteins. G proteins can be grouped into two families, hererotrimeric, or monomeric, but regardless of this, they can exist in an inactive state, but can be rapidly activated, a process that involves the loss of guanine diphosphate and the gain of guanine triphosphate (GTP). Intrinsic GTPase activity can then rapidly convert the GTP to guanine diphosphate, rendering the protein once again inactive. Obviously, this cycle has to be tightly controlled, and many other proteins, including receptors, are often involved.

A more common mechanism for controlling protein activity is the addition of a phosphate group, a process catalysed by kinases. The addition of a phosphate group may alter the ability of the protein to have further protein–protein interactions, or may change the conformation of the protein and so alter its catalytic activity. The process can be reversed by removal of the phosphate group, catalysed by phosphatases.

Therefore, it can be understood that evolution has put in place a whole series of ways in which protein function can be quickly altered, with that alteration being able to be undone. The question therefore arises: Can redox chemistry, and the presence of redox compounds, be involved in such signaling pathways?

## 3. Redox-Mediated Signal Transduction

The main compounds that are thought to be involved in redox-mediated signal transduction are NO and those that come under the umbrella of ROS, that is, hydrogen peroxide ($H_2O_2$) and superoxide anions ($O_2^{\bullet-}$). Other compounds such as hydroxyl radicals or peroxynitrite ($ONOO^-$) have also been studied, with the latter recently attracting attention as a potential signaling molecule *(6)*.

### 3.1. Redox Molecules as Signals

If NO and ROS are to act as signals they need to have certain characteristics:
- they need to be produced where they are needed;
- they need to be produced rapidly and only when needed;
- they need to be perceived, and their presence acted upon; and
- they need to be removed rapidly, so that the signal is not sustained.

For both NO and ROS, there appears to be no single source of their production that can be pointed at and deemed to be part of a signaling pathway. In mammals, NO is primarily produced by a family of nitric oxide synthase (NOS) enzymes, two of which are constitutively present (endothelial NOS and neuronal NOS), and one of which is inducible (iNOS). They are all flavin- and haem-containing enzymes and use arginine as a substrate, which is converted to citrulline and NO by a short redox pathway. In fact, the enzymes have a high degree of similarity to the P450 system but, here, they have a reductase domain and an oxidase domain as part of the same polypeptide. However, NO can also be produced by other redox enzymes, such as the molybdenum-based xanthine oxidoreductase (otherwise known as xanthine oxidase). In plants, it has been found that another molybdenum-based enzyme, nitrate reductase, produces NO from nitrite. Therefore, to implicate any one source of NO to a particular pathway may not be a straightforward as first anticipated (for a review on NO signaling, *see* **ref.** *7*, for example).

ROS, on the other hand, also is produced from a range of sources and, in fact, ROS can arise from many redox pathways. For example, both mitochondria and chloroplasts are known to produce ROS. In addition to producing NO, the enzyme xanthine oxidoreductase was first studied because it was known to produce ROS, and many organisms have enzymes that are related to the NAPDH oxidase system. NADPH oxidase was first studied in neutrophils in mammalian host defence, but it is now known that many homologues are involved in ROS production for signaling purposes.

By examining the enzymes involved in the production of NO and ROS, one can understand that many are themselves controlled in a tight manner, often by the intracellular $Ca^{2+}$ levels and, therefore, the generation of the relevant redox signaling molecule is well controlled. Furthermore, often the substrates for the reactions are reasonably common, or at least not scarce, materials, such as oxygen, NADPH, nitrite, or arginine. This means that the production of NO or ROS can usually be rapid and controlled, fulfilling requirements of signaling molecules outlined above.

Signaling molecules also need to be removed when no longer needed, and here the very nature of the compounds helps. NO and ROS are inherently unstable. NO is rapidly converted to nitrites and nitrates, whereas superoxide will dismute to $H_2O_2$. Furthermore, cells are replete with antioxidants and antioxidant mechanisms, such as catalase, glutathione, ascorbate, and enzyme systems, to recycle such compounds. Therefore, any ROS produced are very likely to be removed very quickly and, in fact, the rapid removal of ROS and NO has been used an a argument against their involvement in signaling, because it sometimes unclear how the ROS or NO can reach the proteins that they are potentially controlling.

### 3.2. ROS and NO perception

For ROS and NO to be true signals, they need to be perceived by cells, either when they are arriving from the outside, or when they are produced intracellularly. If they are not perceived in some way, there will be no response, and no further signaling will take place.

It is hard to conceive that there are receptors for NO and ROS that function as "classical" receptors. For example, $H_2O_2$ is too small to partake in a ligand–receptor binding event. Therefore, it is more likely that ROS and NO use the nature of their chemistry to have their presence felt.

The most well-characterized mechanism for the action of NO is through its reaction with the enzyme guanylyl cyclase (otherwise referred to as guanylate cyclase), promoting the production of cyclic guanosine monophosphate (cGMP or guanosine 3′,5′-cyclic monophosphate) as depicted in **Fig. 1.1**. The exact target on the cyclase for the NO is the enzyme's haem group, and NO can also have an effect on other haem-containing enzymes. In addition, it has been suggested that some haem-containing enzymes may be involved in modulating NO levels in cells *(8)*. However, both ROS and NO can chemically modify proteins, and the mechanisms involved and the ramifications of such modifications are attracting great interest at the moment.

### 3.2.1. Covalent Modifications Involving ROS and NO

One of the main chemical targets for ROS and NO is the thiol group of the amino acid cysteine. As depicted in **Fig. 1.2**, there are a variety of modifications that are possible on such thiols. Simple oxidation of the thiol in the presence of another not too distant cysteine (in three dimensions) may lead to the formation of a disuphide bridge, a simple –S–S– structure (not shown in **Fig. 1.2**). However, oxidation of a single thiol can lead to the formation of a sulfenic acid group. In some cases, for example, in the tyrosine phosphatase this has been found to rearrange to

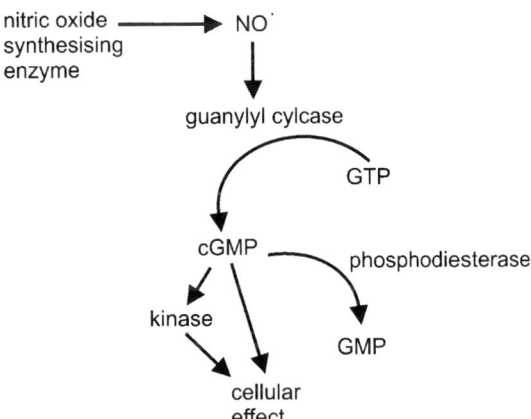

Fig. 1.1. Nitric oxide (NO) often has its affects through the activation of guanylyl cyclase.

form the sulphenyl amide group *(9, 10)*. Further oxidation of the sulfenic acid can lead to the formation of sulfinic acid or further to the sulfonic acid. The more oxidized this thiol becomes, the less likely it is that it can be re-reduced, and so formation of the sulfonic acid is thought to be irreversible, not good for cell signaling, where reversibility is often desired. On the other hand, reaction of NO with the thiol can lead to the formation of –SNO, a process known as S-nitrosylation. The identification of such modifications is the subject of a chapter later in this book (Chap. 15 and *see* **ref**. *11*).

Further thiol modification can also take place with thiols reacting with a common antioxidant found in cells, that is, glutathione. This process of glutathionylation has been shown to be a potential way of controlling a wide range of proteins *(12)*, and further work in this area will no doubt increase the interest in this biochemistry.

Therefore, it can be seen that modification of proteins can be the result of exposure to redox compounds such as ROS and NO. Such modifications would potentially be associated with conformational changes in the protein, with the resultant change of activity, either increasing it or decreasing it. This is analogous to the mechanism of phosphorylation, for which there is no doubt about its importance in cell signaling.

Of course, if NO and ROS, and even antioxidants such as glutathione are all vying to react with thiol groups, there will no doubt be a competition between them. It may be that it is the exact concentrations of ROS, NO, and glutathione that dictate the exact modification that finally takes place. Each type of modification, either

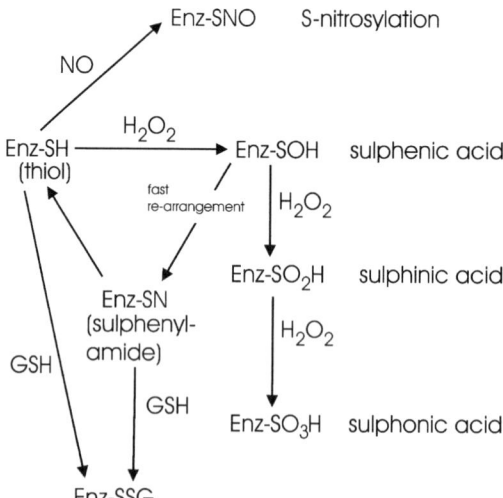

Fig. 1.2. Cysteine thiol groups can partake in a variety of chemistry, perhaps in competition with each other.

S-nitrosylation, oxidation, or glutathionylation, will potentially have a different effect on the activity or functioning of the protein, and so the activity of NOS, ROS producing oxidases and other mechanisms for generating alterations in redox need to be considered in a holistic way.

**3.2.2. Proteins Involved in Perception of ROS and NO**

One of the challenges in redox signal transduction research has been to identify the proteins involved *(13)*. Some proteins known to be involved in signal transduction have been found to be inhibited by $H_2O_2$, for example, tyrosine phosphatase *(14)*, but other proteins also are undoubtedly involved in redox signaling. By the use of a microarray approach in a study of cells after a treatment of oxidative stress, the ethylene receptor ETR1 was found to be important in $H_2O_2$ signaling in *Arabidopsis thaliana*. It was found by the use of mutants that a cysteine residue was important for this functioning (Cys65), suggesting that thiol modification is involved, although this has yet to be confirmed (*see* **ref.** *15* and Chap. 7). In an alternate approach, thiol tagging with fluorescent markers in the presence and absence of ROS revealed several proteins that could potentially be involved in $H_2O_2$ signaling, including glyceraldehyde 3-phosphate dehydrogenase (i.e., GAPDH: *see* **ref.** *16)*, a protein more usually associated with metabolism, in particular the glycolytic pathway.

Transcription factors have also been identified that are controlled in a redox-mediated manner. In *Escherichia coli*, for example, the transcription factor OxyR is activated by $H_2O_2$, leading to a disulfide bond formation between cysteine 199 and cysteine 208 *(17)*.

Therefore, it can be seen that there are a wide variety of proteins that might take part in the perception of redox and redox-active molecules, including ones that are already associated with cell signaling, ones that are involved in other activities such as metabolism, and others that are in direct control of gene expression. The challenge remains to untangle how this all functions and is coordinated.

**3.2.3. Systems Involved in Reversal of Redox Modifications**

Once a cysteine thiol has been oxidised, it needs to be re-reduced if the signaling is to be ablated. If the disulfide bond has been created by ROS reduction, thioredoxins (TRXs) and glutaredoxins may act as protein disulphide reductases *(18, 19)*. TRXs and glutaredoxins may also reoxidize –SOH groups *(20, 21)*. If ROS oxidation has caused the formation of the sulfinic acid group, this can be reduced back to the sulfenic acid group by sulfiredoxins. These ATP-dependent enzymes were first identified in yeast *(22)*. The sulfenic acid group created can then be further reduced by TRX or glutaredoxins, resulting in the regeneration of the thiol, –SH, and so allowing a further round of signaling.

## 4. Conclusion

There are many compounds, from a variety of sources, that can influence the redox status of cells and potentially modify a range of proteins via redox chemistry. Many of these changes to such proteins will result in alterations of their activities and functions; therefore, a greater understanding of how redox is controlled and how changes in redox influence metabolic pathways or the expression of genes is required to fully elucidate how redox chemistry fits into the overall scheme of cell signaling.

The authors and chapters that follow in this book will discuss the influence of aspects of redox chemistry in more depth and discuss some of the key methodologies that may be employed to study this emerging area of research. For a more full text on cell signaling in general and the components involved, *see* **refs**. *23* and *24*.

## References

1. Hancock, J. T. (1997) Superoxide, hydrogen peroxide and nitric oxide as signalling molecules: their production and role in disease. *Br J Biomed Sci* **54**, 38–46.
2. Dröge, W. (2002) Free radicals in physiological control of cell function. *PhysiolRev*, **82**, 47–95.
3. Neill, S. J., Desikan, R., and Hancock, J. T. (2002) Hydrogen peroxide signalling. *Curr Opin Plant Biol*, **5**, 388–395.
4. Colavitti, R., and Finkel, T. (2005) Reactive oxygen species as mediators of cellular senescence. *IUBMB Life*, **57**, 277–281.
5. Vanderauwera, S., Zimmermann, P., Rombauts, S., Vandenbeele, S., Langebartels, C., Gruissem, W., Inzé, D., and Van Breusegem, F. (2005) Genome-wide analysis of hydrogen peroxide-regulated gene expression in *Arabidopsis* reveals a high light-induced transcriptional cluster involved in anthocyanin biosynthesis. *Plant Physiol*, **139**, 806–821.
6. Ullrich, V., and Kissner, R. (2006) Redox signaling: bioinorganic chemistry at its best. *J Inorg Biochem* **100**, 2079–2086.
7. Neill, S. J., Desikan, R., and Hancock, J. T. (2003) Nitric oxide signalling in plants. *New Phytol* **159**, 11–35.
8. Perazzolli, M., Romero-Puertas, M. C., and Delledonne, M. (2006) Modulation of nitric oxide bioactivity by plant haemoglobins. *J Exp Bot* **57**, 479–488.
9. Salmeen, A., Anderson, J. N., Myers, M. P., Meng, T. C., Hinks, J. A., Tonks, N. K., and Barford, D. (2003) Redox regulation of protein tyrosine phosphatase 1B involves a novel sulfenyl-amide intermediate. *Nature* **423**, 769–773.
10. Van Montfort, R. L., Congreve, M., Tisi, D., Carr, R., and Jhoti, H. (2003) Oxidation state of the active-site cysteine in protein tyrosine phosphatase 1B. *Nature* **423**, 773–777.
11. Lindermayr, C., Saalbach, G., and Durner, J. (2005) Proteomic identification of S-nitrosylated proteins in *Arabidopsis*. *Plant Physiol* **137**, 921–930.
12. Dixon, D. P., Skipsey, M., Grundy, N. M., and Edwards, R. (2005) Stress-induced protein S-glutathionylation in *Arabidopsis*. *Plant Physiol* **138**, 2233–2244.
13. Hancock, J., Desikan, R., Harrison, J., Bright, J., Hooley, R., and Neill, S. (2006) Doing the unexpected: proteins involved in hydrogen peroxide perception. *J Exp Bot* **57**, 1711–1718.
14. Cho, S.-H., Lee, C.-C., Ahn, Y., Kim, H., Yang, K.-S., and Lee, S.-R. (2004) Redox regulation of PTEN and protein tyrosine phosphatase in $H_2O_2$-mediated cell signalling. *FEBS Lett* **560**, 7–13.
15. Desikan, R., Hancock, J. T., Bright, J., Harrison, J., Weir, I., Hooley, R., and Neill, S. J.

(2005) A novel role for ETR1: hydrogen peroxide signalling in stomatal guard cells. *Plant Physiol* **137**, 831–834.

16. Hancock, J. T., Henson, D., Nyirenda, M., Desikan, R., Harrison, J., Lewis, L., Hughes, J., and Neill, S. J. (2005) Proteomic identification of glyceraldehyde 3-phosphate dehydrogenase as an inhibitory target of hydrogen peroxide in *Arabidopsis*. *Plant Physiol Biochem* **43**, 828–835.

17. Lee, C., Lee, S. M., Mukhopadhyay, P., Kim, S. J., Lee, S.C., Ahn, W. S., Yu, M. H., Stroz, G., and Ryu, S. E. (2004) Redox regulation of OxyR requires specific disulfide bond formation involving a rapid kinetic reaction path. *Nat Struct Mol Biol* **11**, 1179–1185.

18. Schürmann, P., and Jacquot, J. P. (2000) Plant thioredoxin systems revisited. *Annu Rev Plant Phys* **51**, 371–400.

19. Lemaire, S.D. (2004) The glutaredoxin family in oxygenic photosynthetic organisms. *Photosynth Res*, **79**, 305–318.

20. Collin, V., Lankemeyer, P., Miginiac-Maslow, M., Hirasawa, M., Knaff, D. B., Dietz, K. J., and Issakidis-Bourguet, E. (2004) Characterization of plastidial thioredoxins belonging to the new y-type. *Plant Physiol* **136**, 4088–4095.

21. Rouhier, N., Gelhaye, E., Sautiere, P. E., Brun, A., Laurent, P., Tagu, D., Gerard, J., De Fay, E., Meyer, Y., and Jacquot, J. P. (2001) Isolation and characterization of a new peroxiredoxin from poplar sieve tubes that uses either glutaredoxin or thioredoxin as a proton donor. *Plant Physiol* **127**, 1299–1309.

22. Biteau, B., Labarre, J., and Toledano, M. B. (2003) ATP-dependent reduction of cysteine-sulphinic acid by *S. cerevisiae* sulphiredoxin. *Nature*, **425**, 980–984.

23. Hancock, J. T. (2005) *Cell Signalling*, 2nd ed., Oxford University Press, Oxford..

24. Hancock, J. T. (2003) The principles of cell signalling, in *On Growth, Form and Computers*, (Kumar S., Bentley, P. J., eds.), Academic, London. 64–81.

# Chapter 2

## The Measurement of Nitric Oxide and Its Metabolites in Biological Samples by Ozone-Based Chemiluminescence

**Andrew G. Pinder, Stephen C. Rogers, Afshin Khalatbari, Thomas E. Ingram, and Philip E. James**

### Abstract

A plethora of publications on techniques and methodologies for measuring nitric oxide (NO) or reaction products of NO (NO metabolites) has served in recent years to complicate and confuse the majority of researchers interested in this field. Here, we provide a practical approach and summarize the key issues and corresponding solutions regarding quantification with the use of ozone-based chemiluminescence, which is the most accurate, sensitive, and widely used NO detection method. We have drawn on the vast experience of leaders in the field to produce this consensus, but the views and implications presented herein represent our own, and we limit our advice to those techniques with which we have direct experience. Hopefully, this guide will allow authors to make more informed decisions regarding NO metabolite measurement methodology, without the need for each subsequent group to rediscover previously observed advantages and pitfalls.

**Keywords:** Nitric oxide, nitrite, nitrate, nitrosothiols, ozone-based chemiluminescence.

### 1. Introduction

The involvement of nitric oxide (NO) in a vast number of signaling pathways has created a need to accurately measure this free radical species in a variety of biological systems. However, because of the highly reactive nature of free NO, its measurement is practically difficult and extremely complex at best. Therefore, researchers have focused on the quantification of the reaction products of NO (NO metabolites). Particular emphasis has been made on the measurement of nitrate, nitrite, and protein-bound

John T. Hancock (ed.), *Methods in Molecular Biology, Redox-Mediated Signal Transduction, vol. 476*
© 2008 Humana Press, a part of Springer Science + Business Media, Totowa, NJ
DOI: 10.1007/978-1-59745-129-1_2

NO species to obtain information regarding the in vivo production, consumption, and bioavailability of NO.

Ozone-based chemiluminescence (OBC) is generally recognized as the most accurate and sensitive technique available in which to measure NO. However, to directly measure metabolite NO, the NO must either be cleaved from the species of interest or the metabolite must be reduced back to NO. Several methods exist to achieve this state, primarily photolysis or the use of chemical cleavage reagents. However, researchers can be confused by the wealth of literature filled with speculation and counterargument regarding specificity. Over a number of years we have learned the hard way, undertaking the necessary experiments to identify key issues and confounding factors. With considerable success, we (and others) have gained an in-depth understanding and herein provide a useful and practical guide to NO metabolite measurement using chemical cleavage reagents linked to OBC.

## 2. General Methodology

### 2.1. Materials

1. High-performance liquid chromatography (HPLC)-grade water (Fisher Scientific, Leicestershire, UK).
2. Glacial acetic acid (Fisher Scientific UK).
3. Hydrochloric acid (Fisher Scientific UK).
4. Potassium iodide (KI: Sigma-Aldrich Co., Ltd., Dorset UK).
5. Iodine ($I_2$: Sigma-Aldrich).
6. Cuprous chloride (CuCl: Sigma-Aldrich).
7. l-Cysteine (CSH: Sigma-Aldrich).
8. Vanadium chloride ($VCl_3$: Sigma-Aldrich).
9. Antifoam 204 organic (Sigma-Aldrich).
10. Potassium hexacyanoferrate ($K_3Fe^{III}(CN)_6$: Sigma-Aldrich).
11. Sulphanilamide (Sigma-Aldrich).
12. Mercury chloride ($HgCl_2$: Sigma-Aldrich).
13. *N*-ethyl maleimide (NEM: Sigma Aldrich).

### 2.2. Nitric Oxide Analysis by Ozone-Based Chemiluminescence (OBC)

OBC is a measurement technique that exploits the luminescent nature of specific chemical reactions and can be used to quantify NO. As outlined previously in this chapter, before NO (or metabolite NO) can be measured, it must firstly be cleaved from its parent compound(s) or reduced back to its radical form. This can be achieved by the use of a number of chemical reagents, which will be discussed in more detail later. NO is subsequently carried in an inert gas stream (e.g., nitrogen [$N_2$] or argon [Ar] at a constant flow rate ~100–150 cm³/min) to the nitric oxide analyser (NOA).

The NOA uses oxygen to generate ozone ($O_3$) in its reaction cell. NO entering the NOA reacts with $O_3$ to form nitrogen dioxide ($NO_2$) and oxygen ($O_2$). However, a proportion of the $NO_2$ is formed in an electrically excited state ($NO_2^*$). Electrons in this state are unstable and, as they return to their original ground state, they release their excess energy as a photon. Released photons are focused via a low-pass filter lens (<900 nm wavelength) into a photomultiplier tube, which amplifies the signal to give an accurate, recordable millivolt signal.

In our laboratory, we currently use a Sievers NOA 280i (Analytix, Durham, UK). It must be acknowledged that there is variation in sensitivity between machines, and considerable differences between manufacturers of NOA exist. It is therefore worthwhile investing some time in testing actual analysers if considering a purchase.

**2.3. Experimental Setup**

The inert carrier gas ($N_2$) flows through a purge vessel that contains the cleavage reagent (**Fig. 2.1**) and then through a sodium hydroxide (25 mL of 1 $N$ NaOH) trap and solvent filter (Whatman solvent IFD) placed in-line before the NOA. The NaOH and solvent filter serve to protect the NOA reaction cell from damage by hot acid vapor. The NaOH also ensures that N-oxide contaminants are not converted to NO and erroneously measured.

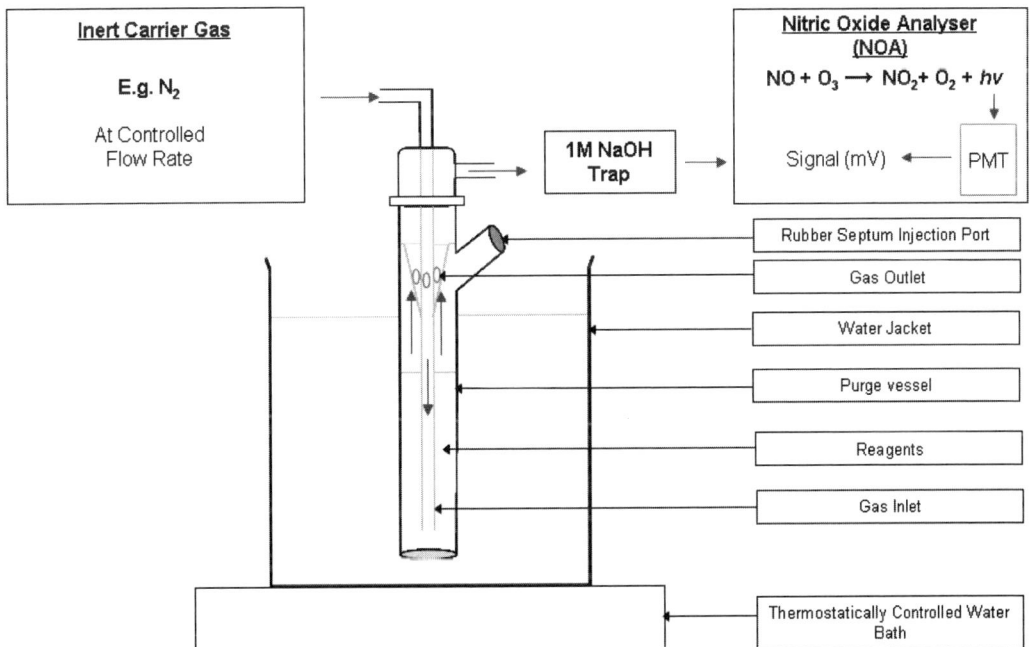

Fig. 2.1. Apparatus used for ozone-based chemiluminescence.

Commercially available purge vessels are provided with pressurized fittings and connectors. Depending on the specific application, they tend to have large reagent volumes that are ideal for multiple sample injections. However, they may not be ideal for certain measurements (for example, NO linked to hemoglobin), where it is essential that the reagent is changed between samples. This is the result of the ability of hemoglobin to scavenge NO within the purge vessel reagent (*see* **Sect. 2.5.**).

In our laboratory, we have opted for custom-designed and -built glassware. This has led to the creation of an interchangeable reagent vessel (~10 mL maximum volume) with side arm injection port and a T-shaped gas purging component, which forms a gas-tight fitting with the reagent vessel. Our setup allows for the quick replacement of the reagent and easy washing of the glassware (**Fig. 2.1**).

For different experimental setups, various temperatures are required. It is therefore essential to accurately maintain the temperature of the purge vessel. We achieve this by using a water bath on a thermostatically controlled hotplate (IKA® WERKE). Samples should be injected into the purge vessel with a glass Hamilton (Fisher) syringe through a rubber septum injection port on the side of the purge vessel.

### 2.4. Calibration

Calibration should be performed on a daily basis to allow for day-to-day variations in machine sensitivity. The standard used for the calibration curve depends on the sample to be measured and the cleavage reagent used.

1. In general, nitrite or *S*-nitrosoglutathione (GSNO) is used as the standard.

2. Prepare standards of 62.5, 125, 250, 500, and 1000 n*M* in HPLC-grade ultrapure water of the volume to be used.

3. Inject a known volume for analysis (usually 100/200 µL) to generate a standard curve of NO (**Fig. 2.2**).

Alternatively, an NO donor (such as NONOates available from Axxora) can be injected into water. In this case, the half-life (t½) for NO release is critical. MAHMANOnoate has a t½ of 2–3 min and is stable at pH 9. It can be conveniently stored on ice and quickly made up in water (pH 7) for injection and accurate production of NO in the purge vessel. It is extremely difficult, if not impossible, to synthesize biological standards that contain NO in physiological proportions, particularly in the case of blood-borne metabolite species.

4. Plot a standard curve of peak area under curve (AUC) against NO (either concentration or molar amount).

5. To generate a straight line relationship, the area under curve of the water injected is removed from the other NO standards to account for water contaminants.

**A**

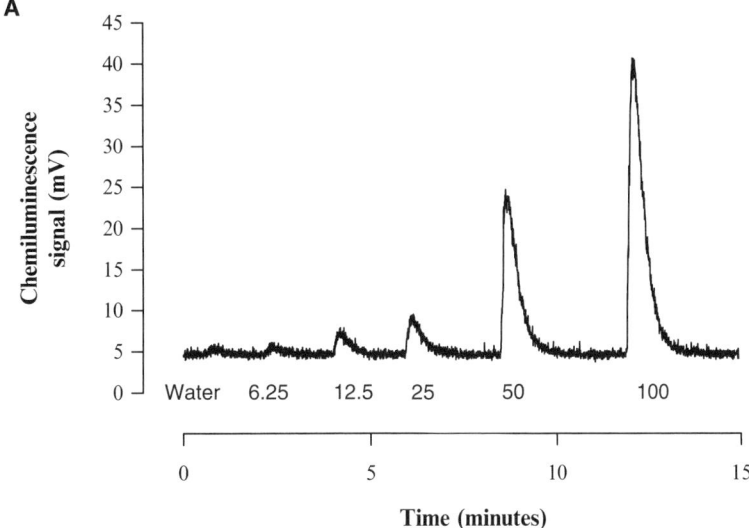

**B**

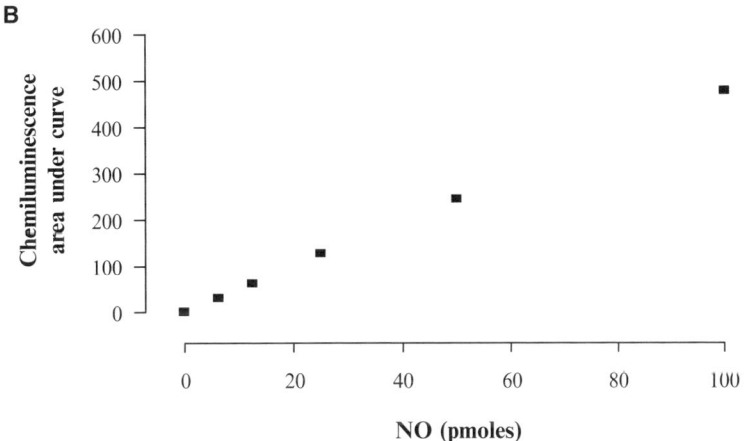

Fig. 2.2. Typical standard curve of nitrite injected into tri-iodide (8 mL) in conjunction with ozone-based chemiluminescence **(A)**. Correlation between NO injected and area under curve **(B)**. Below each peak on the trace is the amount of nitrite injected in pico-moles. Values presented as mean ± SEM ($n = 8$ curves on separate days).

6. After a line of best fit is produced, this relationship is then used to calculate NO from the area under curve of the sample.

*2.5. Analysis of Data*

The AUC typically is measured because it provides the total NO release/detected rather than a time course for NO release. The presence of certain biological matrices within the purge vessel will likely affect the "detectability" of the released NO by the NOA. In principle, this should only serve to broaden the signal detected so long as all the NO is ultimately released by the cleavage reagent from the subsequent products formed (this means that although signal peak height may be influenced the AUC should remain unchanged). This is particularly important with regard to the measurement of NO in samples, where there is a high concentration of heme groups available to rebind the released NO.

An example that we have studied in great detail in the tri-iodide cleavage reagent is NO linked to hemoglobin. At low hemoglobin concentrations, relative to NO (high NO-hemoglobin ratio), the hemoglobin in the reaction chamber simply broadens the NO signal (signal height is significantly reduced but the area under curve remains unaffected: *see* **Fig. 2.3A**).

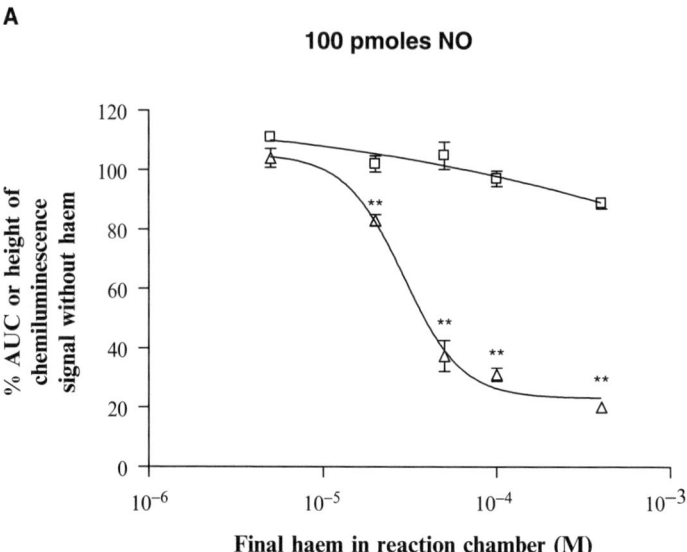

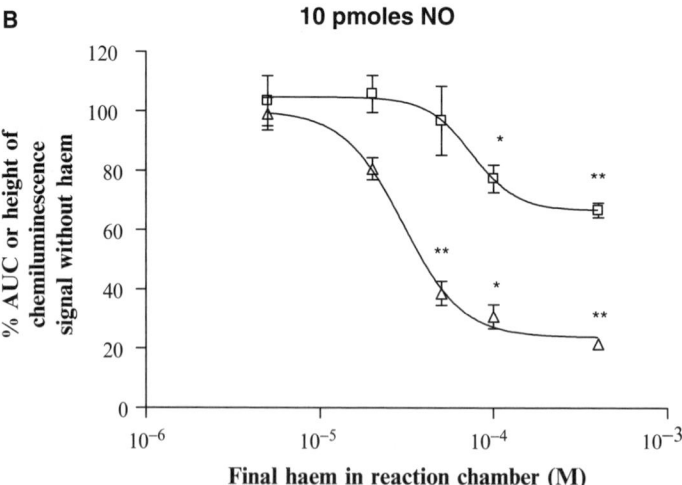

Fig. 2.3. Dose-dependent heme nitric oxide (NO) autocapture. Effect of heme NO scavenging in tri-iodide on signal height and area under curve analysis in the presence of 100 pmol **(A)** and 10 pmol NO **(B)**. Squares represent area under curve analysis, triangles represent signal height. Values are presented as mean ± SEM ($n$ = 4–7 for each concentration). *$p$ < 0.05, **$p$ < 0.01 in relation to 100% values; One-way ANOVA.

However, at high hemoglobin concentrations (and especially where NO-hemoglobin ratio is in the physiological range of >1:25,000) not only is signal height affected, but the AUC also is underestimated (**Fig.2.3B**). To overcome this problem, two techniques have been independently developed: (a) the use of high-pressure CO as the carrier gas in the NOA set up, and (b) the inclusion of $K_3Fe^{III}(CN)_6$ in the tri-iodide reagent (this latter methodology being specific for the tri-iodide reagent only). Both methods prevent the autocapture of the released NO by hemoglobin while not affecting the reductive capacity of the reagents (this means that the same species are measured in both the modified and unmodified assays). We will return to this in **Section 3.3**.

Improved accuracy of AUC measurement (especially for a weak signal) can be gained by exporting the recorded data from the NOA data collection program to a simulation package. We, and others, have found *Origin* to be particularly useful, where by adjacent signal averaging and peak analysis (enabling user placement of peak start and end) can be implemented to improve the signal-to-noise ratio, resulting in a more accurate determination of AUC and improved repeatability of measurement (**Fig. 2.4**). We have also found other software packages to be effective (e.g., *Sigmaplot*) but not as user-friendly.

### 2.6. Step-by-Step Setup

1. Switch on gas supplies to NOA and allow the machine to cool down to set temperature. Inside the machine, the PMT is cooled to less than $-10°C$ (*see* **Note 1**).

2. A purge vessel loaded with the appropriate volume of reagent can then be loaded on to the system. The NOA can then be started; a vacuum pump should cut in to draw sample gas through (*see* **Note 2**).

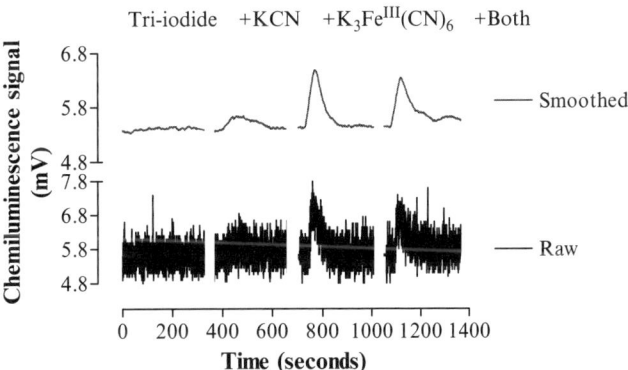

Fig. 2.4. *Measurement* of red blood cell lysate in different reaction mixtures. Reagents include tri-iodide alone; tri-iodide plus potassium cyanide (KCN); tri-iodide plus potassium ferricyanide ($K_3Fe^{III}(CN)_6$), and tri-iodide plus both cyanides. Example traces of chemiluminescence signals and the same data smoothed for analysis with Origin (version 7.0) are shown.

3. Complete a standard curve once signal is at baseline.

4. After completion of a standard curve, change the purge vessel to begin sample testing (*see* **Note 3**).

It is vital that the vacuum has stopped; otherwise, an imbalance in the positive purge gas pressure, back pressure caused by filters, and the negative vacuum pressure can cause the reagent/sodium hydroxide trap to be drawn back through the system as the apparatus is dismantled.

## 3. Cleavage Reagents and Specific Applications

### 3.1. Glacial Acetic Acid

Where measurement of nitrite is the primary aim, sample injection into acetic acid alone in an oxygen-free environment produces NO according to:

$$3HNO_2 \rightleftharpoons 2NO + H^+ + NO_3^- + H_2O$$

From nitrite standards, broad peaks (shorter and longer) are observed in acid alone compared with more-powerful reductive reagents. Acid alone is relatively specific for nitrite detection because it is not strong enough to reduce nitrate to NO. Also, S-nitrosothiols (RSNO; at least GSNO and albumin-SNO) is undetectable and considered stable to acid treatment, although in some cases (e.g., HbSNO), the activation energies of thiol groups is reported to be completely different, implying the same standards and techniques may not be completely appropriate. Examples in which direct injection into acid at 50°C detects nitrite efficiently include supernatant from cell culture experiments or assessment in saline or other medium.

### 3.2. Glacial Acetic Acid and Potassium Iodide

Originally, a reaction mix comprising potassium iodide in glacial acetic acidic was used to determine nitrite in biological fluids. However, this reagent was subsequently demonstrated to detect NO from other biological metabolite species (including *S*-nitrosothiols). The detection of RSNO was found to vary widely *(1)* because of the uncontrolled formation of free iodine from the following reaction:

$$2HNO_2 + 2I^- + 2H^+ \rightarrow 2NO + I_2 + 2H_2O$$

The formation of free iodine ($I_2$) allowed the production of tri-iodide ($I_3$), a reagent capable of detecting RSNO and *N*-nitrosamine (RNNO) in addition to nitrite. Herein lay the major pitfall: The reductive power of the glacial acetic acid/KI reagent was not consistent in its yield return from RSNO.

The use of KI/glacial acetic acid is only effective after the addition of KI to deoxygenated acid with the subsequent maintenance of an $O_2$-free environment. If oxygen is introduced, even in small quantities, the formation of $I_3$ results. Under such conditions, there is no difference between the reduction of nitrite to NO in KI/glacial acetic acid compared with tri-iodide.

### 3.3. Tri-iodide

This reagent comprises the following ingredients:

1. 70 mL of glacial acetic acid.
2. 650 mg of iodine ($I_2$).
3. 20 mL of HPLC-grade water.
4. 1 g of KI.

The rationale behind the development of "tri-iodide" was to provide a reductive reagent that was capable of detecting other biological NO metabolite species (i.e., *S*-nitrosothiols) in addition to nitrite. This was achieved by the addition of free iodine to the glacial acetic acid/KI reagent to saturate the solution thus forcing the production of $I_3^-$ in abundance *(1)*.

$$I_2 + I^- \rightarrow I_3^-$$

$$I_3^- + 2RSNO \rightarrow 3I^- + RS\text{-}SR + 2NO^+$$

$$2NO^+ + 2I^- + 2H^+ \rightarrow 2NO + I_2 + 2H_2O$$

To make this reagent:

1. First, dissolve KI in 20 mL of HPLC water.
2. Add iodine to 70 mL of acetic acid.
3. Combine and stir the two solutions for 30 min until all the iodine has dissolved.

For biological samples (e.g., plasma) 5 mL of reductive reagent is loaded into the purge vessel with 20 µL of antifoam. Two to three samples (100–200 µL) can be tested before changing the reagent.

The measurement of red blood cell/hemoglobin-bound NO requires the use of a modified tri-iodide reagent *(2)*. The modified tri-iodide reagent contains potassium ferricyanide to block hemoglobin from "auto-capturing" NO (*see* **Sect. 2.5** and **Fig. 2.4**).

For the modified reagent:

1. Add 800 µL of a 250 m$M$ solution of potassium ferricyanide (823 mg into 10 mL of HPLC-grade water) to 7.2 mL of tri-iodide (25 m$M$ final) with 20 µL of antifoam.

With regard to this modified reagent, two essential aspects should be considered. First, the timing of $K_3Fe^{III}(CN)_6$ addition to tri-iodide should be kept the same for all samples because the reductive capacity of the reagent can change if left for a long period (>30 min). We add $K_3Fe^{III}(CN)_6$ to the reagent 10 min before injection of the sample to allow time for the signal from contaminant nitrite in the $K_3Fe^{III}(CN)_6$ to decay and the temperature of the reagent to reach 50°C, thus providing a stable baseline OBC signal. Second, only one red blood cell (RBC) sample should be injected per reagent mix to limit the amount of hemoglobin in the reaction chamber (minimizing NO–hemoglobin interaction), which also ensures that the reductive potential of the reagent is the same for all samples (*see* **Note 4**).

The fact that the sample is plunged into hot acid and that most biological NO species can potentially be detected in $I_3$ (as a result of its enhanced reductive potential) can cause trouble when interpreting results. A further issue is the fact that the mechanism of precisely how $I_3^-$ works has yet to be elucidated. Nevertheless, $I_3^-$ is probably the most widely used reagent in conjunction with chemical pretreatments that enable the putative measurement of individual metabolite species. Tri-iodide has been validated against standards of $NO_2^-$, GSNO, albumin-SNO, and Hb/RBC standards saturated with NO.

### 3.4. Vanadium Chloride

This reagent comprises the following ingredients:
1. 785 mg of $VCl_3$.
2. 20 mL of HPLC-grade water.
3. 80 mL of 1 *M* hydrochloric acid (HCl).
   - Prepare a saturated solution of $VCl_3$.
   - Stir for 10 min.
   - After stirring filter the solution, it should appear turquoise blue in color.
   - Load 8 mL of the reagent into the purge vessel with 20 µL of antifoam.

The $VCl_3$ reagent can be used to measure nitrate ($NO_3^-$) in addition to those species observed with tri-iodide. Nitrate levels in biological samples tend to be orders of magnitude greater than nitrite levels.

### 3.5. CuCl/CSH

This reagent comprises the following ingredients:
1. 47.25 mg of CSH.
2. 0.1 mg of CuCl (made as 39.59 mg in 10 mL of water to create a 40 m*M* stock solution; then dilute 1/10 to give 4 m*M*, then add 10 mL below to create a 100 µ*M* final concentration).
3. 390 mL of HPLC-grade water.
   - Prepare the reagent fresh, daily.
   - Mix well before pH correction to pH 7 with sodium hydroxide.

The CuCl/CSH reagent was developed specifically to quantify RSNO compounds *(3)*. The reagent takes advantage of the trans-nitrosation reactions that occur between biological nitrosothiols and the cleavage of NO from RSNO compounds by copper ions. The neutrality of this reagent ensures that other metabolites such as nitrates and nitrites remain undetected, guaranteeing specificity.

Cu/CSH is useful because (like photolysis) it is reported to be specific for cleavage of the S-NO group, although the efficiency with which this occurs (30–60%) may depend on other factors (such as varying concentration of reactants, pH):

$$RS\text{-}NO + Cu^+ + H^+ \rightarrow RSH + NO + Cu^{2+}$$

$$2RSH + 2Cu^{2+} \rightarrow RS\text{-}SR + 2Cu^+ + 2H^+$$

Inclusion of cysteine improves NO yield to around 80%. There remains a modest lack of reproducibility, which has been attributed to the fact that decomposition of nitrosothiols under these conditions can occur by several pathways, some of which are described by the equations above. In addition, the presence of an oxidizer will decompose nitrosothiols to $NO^+$ that in turn hydrolyses to nitrite. A more recent and important development for use with RBC samples is the simultaneous perfusion of the Cu/CSH reagent with CO at high reaction cell pressure to prevent auto-capture of the released NO by free Hb in the reaction chamber (*see* **ref.** *4* and **Fig. 2.5**).

***3.6. Photolysis***    This chapter has so far considered chemical cleavage reagents, but it is important to recognize that intense light at specific wavelengths can also be used to cleave NO from proteins and can be used in conjunction with OBC. Photolysis as a cleavage method has not been used in our laboratory but is used successfully by a select number of groups in the field and is worthy of consideration *(5, 6)*. Samples are vaporized and irradiated at a wavelength of 340 nm, which is reported to specifically cleave RNNO, RONO, RSNO, and RMNO (where M is a metal). Nitrite is relatively resistant to photolysis, making this method ideal for the analysis of species that are perhaps not as accessible using chemical cleavage. In the instance of RSNO measurements, treatment with mercuric chloride can be used to validate the origin of the signal. A basic photolysis system is not generally available commercially and a bespoke setup can come at considerable cost.

**Table 2.1**
**Summary of Different Reagents and the NO Metabolite Species Detectable**

| | Cuprous chloride/ cysteine | Tri-iodide | Vanadium chloride |
|---|---|---|---|
| Reagent temperature | 50 (±1)°C | 50 (±1)°C | 85 (±1)°C |
| Species cleaved or reduced | RSNO | $NO_2^-$, RSNO, RNNO, RONO, RMNO (where M is a metal) | $NO_3^-$, $NO_2^-$, RSNO, RNNO, RONO, RMNO (where M is a metal) |

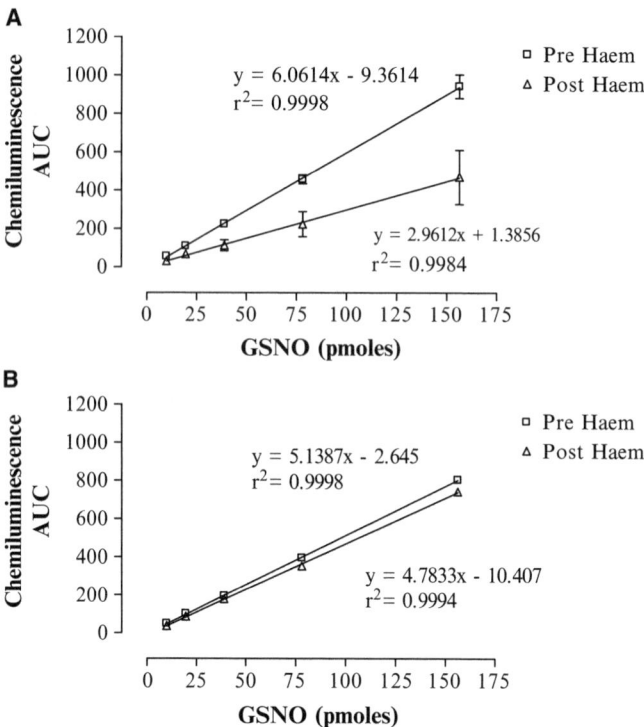

Fig. 2.5. Inhibitory effect of carbon monoxide on nitric oxide (NO) autocapture by cell free heme in the CuCl/CSH reagent. Purge gas flow of CO with normal system reaction cell pressure **(A)** or a positive pressure of 120 mmHg **(B)** to increase effective CO blocking of free Haem (1 μM final heme concentration). Values presented as mean ± SEM (*n* = 3 for each concentration).

Samples also require extensive processing before testing which can lead to contamination issues.

**3.7. Pretreatments**

Notwithstanding the other issues listed here, the various pretreatment regimens have been well publicized and have been used in conjunction with both photolysis and $I_3^-$ linked to OBC (**Table 2.2**). In brief, there are four to consider; mercuric chloride will selectively remove the S-NO component, sulfanilamide (in HCl) will remove $NO_2^-$ over a short incubation time, NEM will prevent trans-nitrosation and in vitro chemistry involving free thiols, and $K_3FeIII(CN)_6$ is deemed to both convert heme-linked NO to $NO_2^-$ and stabilize SNO. In general, these work well with standards.

The mechanisms by which these pretreatments act are not fully understood, but some mechanisms have been proposed. In the presence of acidified sulfanilamide (Ar-NH$_2$), nitrite is reduced to NO$^+$ (reaction A below) which, in turn is believed to react with sulfanilamide (reaction C). Reacted sulfanilamide ultimately forms a stable diazonium salt (reaction D). Mercuric chloride cleaves NO$^+$ from RSNO compounds (reaction B) which

**Table 2.2**
**Summary of Pretreatments Used to Differentiate Between NO Metabolite Species**

| Treatment | Concentration | Use |
|---|---|---|
| Acidified sulphanilamide | 290 m$M$ stock (500 mg in 10 mL of 1 $M$ HCl) Add 1 part acidified sulfanilamide to 9 parts sample | Renders nitrite undetectable by OBC. Requires minimum 15-min incubation in the dark |
| Mercuric chloride | 50 m$M$ stock (67.9 mg in 5 mL of HPLC grade water). Add 1 part to 9 parts sample | Cleaves NO from RSNO compounds |
| NEM | 100 m$M$ stock (62.56 mg in 5 mL of HPLC-grade water) Add 1 part to 9 parts sample | Blocks free thiols to prevent formation of additional RSNO compounds |
| FeCN | 250 m$M$ stock (823 mg in 10 mL of HPLC-grade water) Add 1 part to 9 parts sample | Blocks hemoglobin-NO formation and reported to stabilize hemoglobin-SNO |

in the presence of sulphanilamide will be rendered undetectable to OBC (reactions C and D).

$$NO_2^- + H^+ \leftrightarrows NO^+ + OH^- \leftrightarrows HONO \qquad [A]$$

$$RSNO + Hg_2^+ \rightarrow Hg(RS)_2 + NO^+ \qquad [B]$$

$$NO^+ + Ar\text{-}NH_2 \rightarrow Ar\text{-}N_2^+ + H_2O \qquad [C]$$

$$Ar^{\cdot+} + Ar\text{-}N_2^+ \rightarrow Ar\text{-}N{\equiv}N\text{-}Ar^- \qquad [D]$$

A very important point to consider is the fact that pretreatment is likely to either produce further NO-like species or at least perhaps create an imbalance in the NO metabolite profile of the original sample. Special care must therefore be taken when comparing treated versus untreated samples.

Sample separation or purification also needs careful consideration. Protein removal by the use of filters or by precipitation adds several potential stages for contamination. Column separation (such as with G25) is not effective in removing all $NO_2^-$ or $K_3FeIII(CN)_6/KCN$ from the sample; the latter can in turn give rise to an OBC signal when the sample is injected. In fact, the addition of diluent or washing the sample is a tricky business, not least because this introduces $NO_2^-$ to the sample, which cannot be accounted for by simple removal of unreacted $NO_2^-$ at a later point (such as with sulfanilamide/HCl) before measurement. Finally, even pure chemicals themselves (i.e., "pretreatments") are likely to contain trace amounts of $NO_2^-$.

# 4. Contamination

Nitrite represents the single most significant cause of unwanted OBC signal. Despite careful laboratory practice, nitrite can contaminate samples from a number of sources:

1. *Buffers and water.* Even the purer, high-quality reagents and water contain nitrite contamination. HPLC-grade water (Fisher) or water for injection (Braun) were found to contain the least contamination in our laboratory's experience. However, once open, these waters also accumulate further nitrite contamination over time. This contamination is unavoidable, although it can be minimized by good practices and accounted for by inclusion of a water injection in the standards, which are made up and tested on a daily basis.

2. *Plasticware, glassware, and apparatus.* All disposable and reusable apparatus will contain considerable nitrite contamination that can be minimized by repeat washing. Plasticware in particular can vary vastly depending on both source and batch. The use of filters and columns in this respect should be avoided. Glassware and syringes should be cleaned extensively (e.g., overnight in Decon®, followed by extensive rinsing with water).

3. *Hamilton syringes.* Syringes that are in regular use should be acid washed followed by repeat neutral washes on a regular basis, as residue builds up in both the needle and in the barrel of the syringe.

4. *Chemicals.* All reagents contain a degree of nitrite contamination.

Despite the challenge of contamination, with care and attention to detail this problem can be overcome.

# 5. Practical Considerations

### 5.1. Freezer Time

The lengthy storage of sample in the freezer tends not to alter total NO metabolite levels when compared with freshly measured samples. However, the apportionment of NO between metabolite species does begin to alter at approximately the 7th day of storage (**Figs. 2.6** and **2.7**). As samples are taken, they should be immediately snap frozen in liquid nitrogen and subsequently stored at −80°C. Upon testing, samples should be rapidly thawed (within 3 min at 37°C) in the dark.

*5.2. Sample Preparation*

Generally, sample preparation should be kept to a minimum because of the aforementioned contamination issues. Experimental conditions may dictate that immediate measurement is not possible, for example, when a time-course of sampling is undertaken. Many of our studies that use human subjects dictate that the samples are snap frozen in liquid nitrogen and kept at −80°C for OBC at a later stage. Whether the sample (fresh or from the freezer) is then allowed to equilibrate at 37°C or pretreated for set incubation times can be critical, especially if comparing parallel samples incubated with versus without pretreatment. There is also the question of time for sample separation, e.g., via a G25 column. The key factor is the chemical stability of the species of interest and the potential for contamination (as discussed in **Section 4**). Some investigators have developed a "stabilizing" cocktail that is introduced to the sample as the sample is drawn. It contains NEM (to bind free thiol and prevent in vitro transnit-

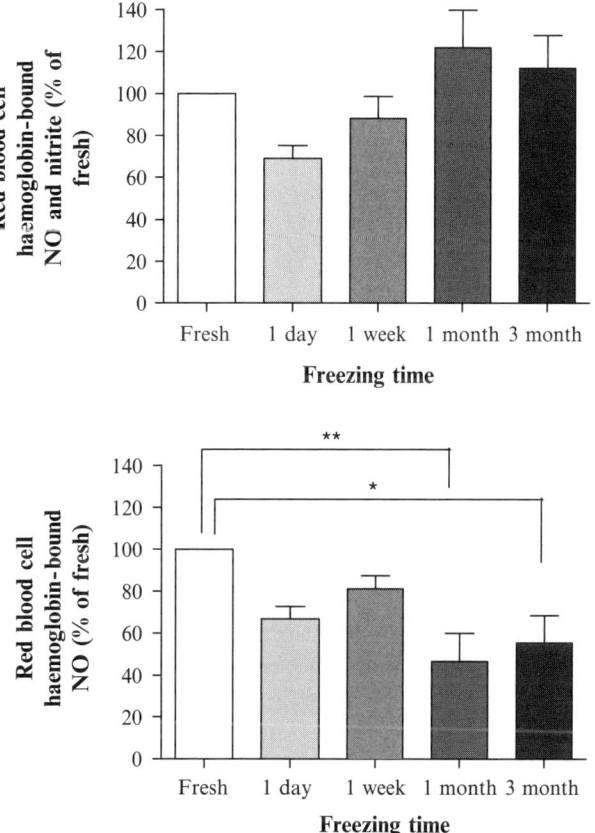

Fig. 2.6. Effect of sample freezing on red blood cell-associated nitrite and hemoglobin-bound nitric oxide *(top)* and red blood cell hemoglobin-bound NO *(bottom)* measured with tri-iodide with added $K_3Fe^{iii}(CN)_6$ in conjunction with chemiluminescence. Fresh represents snap frozen and immediately measured. Values presented as mean ± SEM ($n = 5$); $*p < 0.05$; $**p < 0.01$. Repeat-measures analysis of variance.

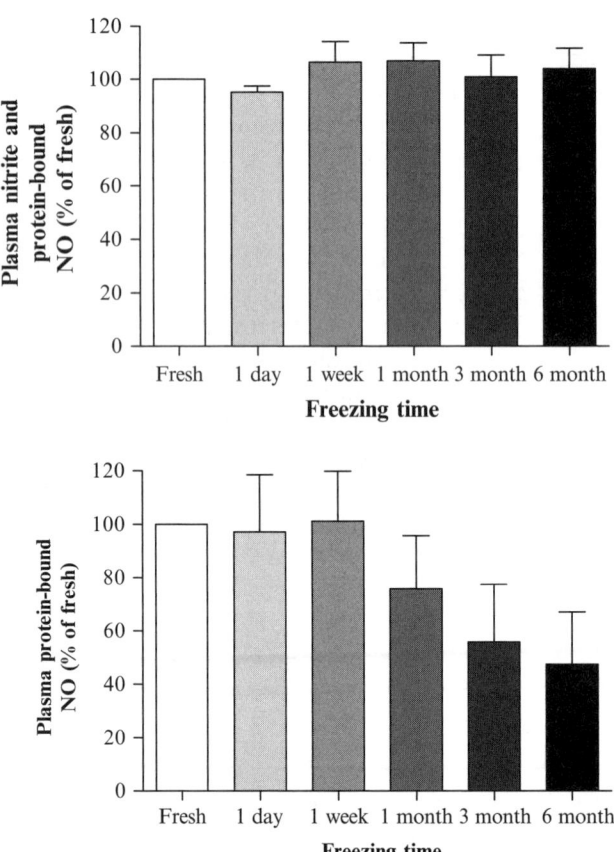

Fig. 2.7. Effect of sample freezing on plasma nitrite and protein-bound nitric oxide levels (top) and plasma protein-bound nitric oxide (bottom) measured using tri-iodide in conjunction with chemiluminescence. Fresh represents snap frozen and immediately measured. Values presented as mean ± SEM ($n = 5$)

rosation) and $K_3FeIII(CN)_6$ (to stabilize the SNO in the sample). This cocktail is also touted to prevent the recapture of NO by free heme in the reaction vessel during the measurement, a critical practical consideration described above.

**5.3. Changing Oxygen/pH**

Alterations of pH can lead to in vitro chemistry between NO metabolite species. Even a minor decrease in pH below 7 is enough to reduce some nitrite back to NO. Changes in oxygen are also important in the case of RBC NO metabolite measurements. A switch in the conformational/oxygenation state of hemoglobin is believed to alter the location of NO within the hemoglobin structure (e.g., NO bound to the oxygen binding site [HbNO] or NO bound to a cysteine residue of the hemoglobin β chain [SNO-Hb]).

## 6. Concluding Remarks—NO Consensus?

There is consensus that OBC is the most sensitive and robust method of detecting fentomol concentrations of NO. The critical consideration is the means to release this NO from the sample. In our experience it is essential to keep it simple. This implies that the less one has to manipulate or add to the sample, the better. Thus, we rely heavily on our *total* measure as being minimally perturbed, in that the samples can be measured directly either fresh or from the freezer. After we have established this *total measure*, it then becomes a secondary question of our assessment of the various NO components with the use of various pretreatments, with the overriding caveat that the sum of the individual components must equal the *total measure*.

We do not invoke any one methodology as the assay of choice. However, we objectively provide a rationale for the procedure(s) we have developed in response to appreciating the pitfalls. Extreme care and consideration must be taken, but the results are robust and point to a real evaluation of NO metabolite levels and to their possible roles.

## 7. Notes

1. Fluctuations in this temperature can alter both baseline and sensitivity, so they should be monitored in a hot laboratory. The flow of gases should also be checked because individual machines and applications will require different flows, but the NOA always requires sufficient $O_2$ supply pressure to form $O_3$.

2. The purge vessel requires some time to come up to temperature and for the machine to find a steady baseline.

3. To change the purge vessel, the NOA must first be set to purge the system. This process takes a few minutes and, when it has finished, the vacuum pump will cut out.

4. Care should be taken to introduce samples directly into the reductive reagent because samples can dry on the inside of the glass purge vessel resulting in an underestimation of NO content.

5. May require warming to dissolve.

## References

1. Samouliov, A., and Zweier, J. L. (1998) Development of chemiluminescence-based methods for specific quantitation of nitrosylated thiols. *Anal Biochem* **258**, 322–330.

2. Rogers, S. C., Khalatbari, A., Gapper, P. W., Frenneaux, M. P., and James, P. E. (2005). Detection of human red blood cell-bound nitric oxide. *J. Biol. Chem.* **280**, 26720–26728.

3. Fang, K., Ragsdale, N. V., Carey, R. M., MacDonald, T., and Gaston, B. (1998) Reductive assays for *S*-nitrosothiols: implications for measurements in biological systems. *Biochem Biophys Res Commun* **252**, 535–540.

4. Doctor, A., Platt, R., Sheram, M. L., Eischeid, A., McMahon, T., Maxey, T., Doherty, J., Axelrod, M., Kline, J., Gurka, M., Gow, A., and Gaston, B. (2005) Hemoglobin conformation couples erythrocyte S-nitrosothiol content to $O_2$ gradients. *Proc Natl Acad Sci USA* **102**, 5709–5714.

5. Stamler, J. S., Jaraki, O., Osborne, J., Simon, D. I., Keaney, J., Vita, J., Singel, D., Valeri, C. R., and Loscalzo, J. (1992) Nitric oxide circulates in mammalian plasma primarily as an S-nitroso adduct of serum albumin. *Proc Natl Acad Sci USA* **89**, 7674–7677.

6. McMahon, T. J., Ahearn, G. S., Moya, M. P., Gow, A. J., Huang, Y. C., Luchsinger, B. P., Nudelman, R., Yan, Y., Krichman, A. D., Bashore, T. M., Califf, R. M., Singel, D. J., Piantadosi, C. A., Tapson, V. F., and Stamler, J. S. (2005) A nitric oxide processing defect of red blood cells created by hypoxia: deficiency of S-nitrosohemoglobin in pulmonary hypertension. *Proc Natl Acad Sci USA* **102**, 14801–14806.

# Chapter 3

# Detection and Measurement of Reactive Oxygen Intermediates in Mitochondria and Cells

Matthew Whiteman, Yuktee Dogra, Paul G. Winyard, and Jeffrey S. Armstrong

## Abstract

Reactive oxygen intermediates (ROIs) play a key role in a number of human diseases either by inducing cell death, cellular proliferation, or by acting as mediators in cellular signaling. Therefore, their measurement in vivo and in cell culture is desirable but technically difficult and often troublesome. To address some of the key methodological issues in examining the formation of ROI in cells and mitochondria, this chapter discusses the following: (a) the cellular sources of ROI and their enzymatic removal, (b) common methods used to determine cellular and mitochondrial ROI such as chemiluminescence, electron paramagnetic resonance spectroscopy, fluorescence, and enzymatic techniques, and (c) some common problems associated with these assays and the interpretation of data. We also provide some simple protocols for the estimation of ROI production in cells and mitochondria, and when measuring ROI in cells and mitochondria, we emphasize the need for thorough understanding of results obtained and their interpretation.

**Keywords:** Assay, chemiluminescence, cytochrome c, electron paramagnetic resonance spectroscopy, fluorescence, mitochondria, reactive oxygen intermediates.

## 1. Introduction

Reactive oxygen intermediates (ROIs), including superoxide anion ($O_2^{\bullet-}$), hydroxyl radicals ($^{\bullet}OH$), and hydrogen peroxide ($H_2O_2$), are generated during aerobic metabolism. The mitochondrion is the most common source of ROI production, although there are other important cellular sources, including enzymes such as cytochrome P450 in the endoplasmic reticulum, lipoxygenases, cyclooxygenases, xanthine oxidase, and the

John T. Hancock (ed.), *Methods in Molecular Biology, Redox-Mediated Signal Transduction, vol. 476*
© 2008 Humana Press, a part of Springer Science+Business Media, Totowa, NJ
DOI: 10.1007/978-1-59745-129-1_3

nicotinamide adenine dinucleotide phosphate-oxidase (NADPH) oxidase of phagocytic cells. Approximately 1–3% of the electrons carried by the mitochondrial electron transport chain (ETC) under normal physiological conditions leak out of the pathway and pass directly to oxygen, generating $O_2^{\bullet-}$ *(1)*. Mitochondrial NADH–ubiquinone oxidoreductase (respiratory complex I) is an important source of production especially in the reverse transport mode *(2)*, whereas ubiquinol–cytochrome *c* oxidoreductase or cytochrome $bc_1$ (respiratory complex III) and the coenzyme Q radical generated during the Q cycle is likely to be the most important site of mitochondrial $O_2^{\bullet-}$ production *(1, 3, 4)*. High concentrations of the enzyme manganese superoxide dismutase (MnSOD) in the mitochondrial matrix ensure that basal levels of $O_2^{\bullet-}$ formed during normal electron transport are kept at a bare minimum to limit oxidative damage to mitochondrial matrix proteins involved in the regulation of metabolism, such as the iron–sulfur proteins aconitase and succinate dehydrogenase *(1, 5)*. MnSOD catalyses the dismutation of $O_2^{\bullet-}$ to yield hydrogen peroxide ($H_2O_2$), which is an important signaling molecule generated throughout the animal and plant kingdoms. Under physiological conditions, $H_2O_2$ generated by the dismutation of mitochondrial $O_2^{\bullet-}$ is reduced by a glutathione (GSH)-dependent peroxidase enzyme (GSH/Px) to yield water whereas in the cytosol, $H_2O_2$, generated by peroxisomal β-oxidation of fatty acids, or by the enzyme xanthine oxidase (XO), is decomposed by catalase and GSH/Px, respectively.

However, when $H_2O_2$ is produced in excess it is also a toxic molecule that can react with ferrous iron ($Fe^{2+}$) to form $^{\bullet}OH$ (either via the Fenton reaction, the metal catalyzed Haber–Weiss reaction or with the neutrophil oxidant hypochlorous acid) *(1)*, which is a short-lived species that can react with molecules such as deoxyribonucleic acid (DNA; *see* **Fig. 3.1.** for a schematic representation of cellular and mitochondrial "ROI" production pathways). In the mitochondrion, the cooperative action of the enzymes such as MnSOD and the GSH/GSHPx system ensure that the ROIs ($O_2^{\bullet-}$ and $H_2O_2$) generated during normal aerobic metabolism are kept to nontoxic levels. Although cells possess extensive antioxidant defence systems (extensively reviewed in **ref**. *1*) to combat ROI toxicity, including catalase, SODs, and the GSH and GSH/GSHPx systems, when ROIs overwhelm cellular antioxidant defense systems and redox homeostasis is lost, the result is "oxidative stress," which can lead to cell damage and death.

An overwhelming body of evidence has accumulated that demonstrates that oxidative stress is a contributing factor in the pathogenesis or clinical progression of a wide range of human diseases, such as neurodegenerative disorders (i.e., Huntington's, Parkinson's disease, and Alzheimer's disease), retinal degenerative disorders, AIDS, cancer, chronic inflammatory conditions

## Sources of Reactive Oxygen Intermediates (ROI) and the primary pathways for their 'detoxification'

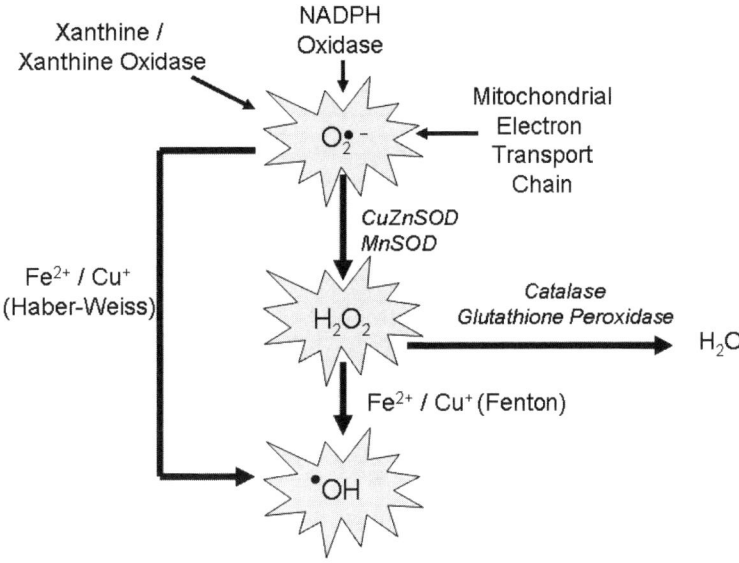

Fig. 3.1. Cellular and mitochondrial sources of ROS production and breakdown. The mitochondrial electron transport chain (ETC), NADPH oxidases, and the xanthine/xanthine oxidase (X/XO) systems are shown as important sources of $O_2^{\bullet -}$. Superoxide dismutases (SODs) enzymatically convert $O_2^{\bullet -}$ to $H_2O_2$, which is broken down to $H_2O$ by catalases and the GSH/GSHPx systems.

such as arthritis, colitis as well as the ageing process in general. Therefore, the ability to make qualitative and quantitative measurements of ROI formation by cells is of particular relevance in addressing disease aetiology, progression, and pathology.

## 2. Methods

### 2.1. Spectrophotometric Methods for the Estimation of ROI

The spectrophotometric methods outlined herein are often the simplest, most widely used, and well-accepted techniques for measurement of extracellular $O_2^{\bullet -}$ production by isolated enzymes, cell homogenates, and isolated phagocytic cells. In general, $O_2^{\bullet -}$ is measured as the SOD-sensitive reduction of a substrate as a result of the nonspecificity of many methods, although this technique does not always guarantee *specificity* for $O_2^{\bullet -}$.

### 2.1.1. Nitro Blue Tetrazolium

Nitro blue tetrazolium (NBT) is a nitro-substituted aromatic compound that can be reduced by $O_2^{\bullet-}$ to the monoformazan, whose formation can be readily monitored by standard bench top spectrophotometers at 550–560 nm. The reaction is a two-step process that proceeds *via* the formation of an intermediate NBT radical, which then undergoes either further reduction or dismutation to the monoformazan *(6, 7)*. NBT detects intracellular $O_2^{\bullet-}$; however, it is less sensitive and specific for $O_2^{\bullet-}$ than fluorimetric assays, such as those that use dihydroethidium. The NBT radical intermediate can also react with molecular oxygen under aerobic conditions and, in doing so, can generate artifactual $O_2^{\bullet-}$ that further reduces NBT *(8)*. Importantly, because the formation of monoformazan is SOD-inhibitable, it illustrates that this technique does not unequivocally confirm that SOD-inhibitable NBT reduction is caused by $O_2^{\bullet-}$ because NBT is also susceptible to reduction by several tissue reductases and has been used to detect cellular enzymes, most notably nitric oxide synthase *(9)*. For these reasons, the detection of $O_2^{\bullet-}$ in biological samples should not exclusively rely on NBT reduction but also should be supplemented with an additional and independent measure of $O_2^{\bullet-}$.

### 2.1.2. Cytochrome c Reduction

Cytochrome *c* reduction is a widely used and well-accepted technique for measurement of $O_2^{\bullet-}$ production by isolated enzymes, cell homogenates, and activated phagocytes. Generally, $O_2^{\bullet-}$ is measured as the SOD-inhibitable reduction of cytochrome *c*, determined in a simple bench top spectrophotometer by the increase in absorbance at 550 nm. Because the cytochrome *c* is cell impermeable, it can be used only to measure extracellular $O_2^{\bullet-}$. There are several precautions when one uses this reaction to detect $O_2^{\bullet-}$. First, cytochrome *c* reduction is also nonspecific for $O_2^{\bullet-}$ and compounds such as ascorbate and glutathione, as well as cellular reductases catalyse cytochrome *c* reduction. In addition, cytochrome *c* can be "reoxidized" by cytochrome oxidase (COX), peroxidases, and a variety of oxidants (including $H_2O_2$ and $ONOO^-$) and, as such, underestimates the true rate of $O_2^{\bullet-}$ production *(10)*. Cyanide ($CN^-$) can be added to the reaction mixture to inhibit COX activity, and ROI scavengers such as catalase will block oxidation by $H_2O_2$ and the addition of urate will prevent oxidation by $ONOO^-$. Alternatively, specificity for $O_2^{\bullet-}$ may be improved by either measuring SOD-inhibitable cytochrome *c* reduction or using acetylated or succinoylated forms of cytochrome *c* which minimise artifactual reduction and oxidation without affecting the ability of $O_2^{\bullet-}$ to reduce cytochrome *c* *(11, 12)*.

### 2.1.3. Aconitase

Aconitase is a citric acid cycle enzyme belonging to the family of dehydrogenases containing iron sulfur (4Fe–4S) centers that catalyse the conversion of citrate to isocitrate. The mitochondrial and the cytosolic forms of aconitase are inactivated by $O_2^{\bullet-}$, and its activity has been proposed to reflect intracellular levels of $O_2^{\bullet-}$ production and with low levels of enzyme activity reflecting high levels of $O_2^{\bullet-}$ (13). Inactivation of the enzyme occurs because of the oxidation of the enzyme and the subsequent loss of Fe from the (4Fe–S) cluster. Unfortunately, other ROIs, $\bullet$NO, and ONOO–, and hypochlorous acid (HOCl) have been shown to inactivate aconitase. As such, the assay is not generally considered specific for $O_2^{\bullet-}$ (14). A serious disadvantage to this method is that because inactivation of the enzyme occurs over several hours or days, it is impossible to intervene with specific scavengers to restore its activity.

### 2.2. Fluorescence Dye Techniques for the Determination of Cellular ROI

Dichlorofluorescin diacetate (DCFH-DA) is taken up by cells and is hydrolyzed to 2′,7′-dichlorofluorescin (DCFH), which is then trapped inside cells. Intracellular DCFH, a nonfluorescent fluorescein analogue, is oxidized by 'ROI' to highly fluorescent 2′,7′-dichlorofluorescein (DCF). DCFH-DA has been quantitatively used to detect ROI produced by PMA-activated phagocytes as well as in cultured cells. Because DCFH is also believed to be oxidized by $H_2O_2$, or $H_2O_2$-generating systems (i.e., glucose oxidase and glucose) and is inhibited by catalase but not by SOD, it was thought that intracellular DCFH oxidation was mediated by $H_2O_2$ (15) and, initially, that the oxidation of DCFH to the fluorescent compound DCF was relatively specific for $H_2O_2$ formation. However, recent studies are emerging to show that this is not the case and that DCF yields fluorescence in response to a variety of ROIs (including $H_2O_2$ and other peroxides, ONOO–, see ref. 16, and hypochlorous acid [HOCl], see ref. 17, after intracellular oxidation mediated by GSH depletion (16, 18–20) and, most importantly, independently of a functional ETC (21), indicating that the ROIs generated under these conditions are not mitochondrial in origin. As a result, *specific scavengers* of 'ROI' must be used with due caution to understand the source of the DCF fluorescence signal.

### 2.2.1. Protocol 1: The Determination of General 'ROI' Formation Using DCFDA and Fluorescence Activated Cell Sorting Analysis

#### 2.2.1.1. Materials

1. DCFH-DA (Molecular Probes, Eugene, OR).
2. Phosphate-buffered saline (PBS), pH 7.4 in 10 m$M$ glucose.
3. Fluorescence-activated cell sorter (FACS) machine or fluorescence plate reader (excitation wavelength 498 nm and emission wavelength 522 nm.
4. Cell culture media (serum free).

#### 2.2.1.2. Method

1. Culture human HL60 cells in RPMI media with 10% fetal calf serum (FCS) and supplements. Cells are maintained in log-phase and kept at a concentration between 2.5 and 5 ×

$10^5$/mL. DCFH-DA is cell permeable because of the diacetate ester. Upon entry into cells, cellular esterases cleave the diacetate to yield the cell-impermeable compound DCFH. Because serum supplements may contain esterases, it is imperative to load cells in serum-free media and centrifuge them to remove culture media before then washing them twice by centrifugation in PBS at 100$g$.

2. Suspend cells in PBS by light vortexing and load with 5–10 µ$M$ DCFDA, incubating them for 15–20 min at 37°C.

3. Wash cells 2× in PBS (centrifuge for 1 min @ 100$g$ and suspend them in PBS containing 10 m$M$ glucose).

4. Because we are performing a viable cell assay, immediate FACS analysis is required, with the use of appropriate software for analysis.

5. The FACS setting should be FL-2 (FITC) with log mode after cell debris have been electronically gated out.

*2.2.2. Protocol 2: The Determination of General 'ROI' Formation With DCFDA and Fluorescence Microscopy: Induction by the Neutrophil Oxidant, Hypochlorous Acid*

2.2.2.1. Materials

1. DCFH-DA (Molecular Probes).

2. PBS, pH 7.4 in 10 m$M$ glucose.

3. Fluorescence microscope with fluorescein filters (i.e., capable of excitation wavelength 498 nm and emission wavelength 522 nm).

4. Cell culture media (serum free).

5. Sodium hypochlorite solution (Sigma).

2.2.2.2. Method

1. Grow human hepatoma HepG2 cells in 24-well plates in minimal essential media, wash with serum-free media, and further incubate in serum-free media containing 1–10 µ$M$ DCFDA or an equivalent volume of vehicle (dimethyl sulfoxide [DMSO]) for 30 min. The final concentration of DCFDA and incubation times may be cell specific and require preliminary investigations for optimization.

2. After this time, wash cells with warm PBS (37°C) and then incubate for 5 min in PBS.

3. Determine hypochlorite concentration at $\lambda$290 nm at pH 12 ($\varepsilon$ = 350/M$^1$/cm) and store on ice diluted in ice cold in ultra high purity water *(17)*. Immediately before use, the hypochlorite solution is diluted in warm PBS and gently poured onto the HepG2 cells. To "quench" hypochlorite, 1 m$M$ methionine solution can be added for 1–2 min before hypochlorite addition.

4. Measure fluorescence as a function of time or hypochlorite addition.

Typical data obtained are shown in **Fig. 3.2.**

**Untreated**    **HOCl**    **HOCl + methionine**

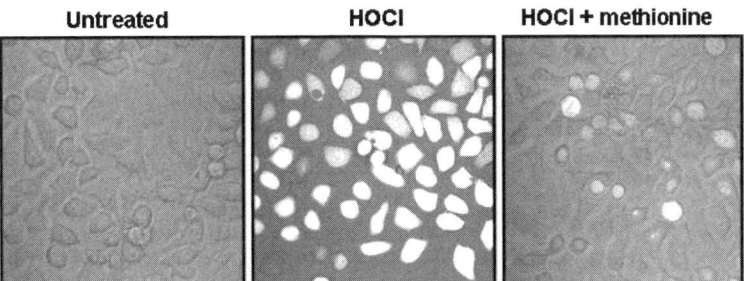

Fig. 3.2. Determination of intracellular 'ROI' produced by human hepatoma HepG2 cells exposed to the neutrophil oxidant, hypochlorite by fluorescence microscopy, and DCFDA.

2.2.2.3. Comments

The use of DCFH for the "specific" measurement of $H_2O_2$ is associated with a number of problems *(22)*. First, it has been reported that the $H_2O_2$-dependent oxidation of DCFH to DCF occurs slowly in the absence of ferrous iron and can be completely inhibited by deferoxamine *(23)*. Second, because peroxidases are capable of inducing DCFH oxidation in the absence of $H_2O_2$ *(24)*, variations in cellular peroxidase activity may influence rates of DCF formation, and cellular fluorescence DCFH itself may also auto-oxidize to form hydrogen peroxide and the reaction of DCFH with peroxidase forms DCF radicals with the subsequent generation of $O_2^{\bullet-}$, suggesting that the use of DCFH to measure ROI may be problematic *(25)*. Third, numerous other substances are capable of directly inducing DCF formation in the absence of $H_2O_2$, including $ONOO^-$ and HOCl. Because of the multiple pathways that can lead to DCF fluorescence, and the inherent uncertainty relating to endogenous versus artifactual oxidant generation, this assay may best be applied as a qualitative marker of *cellular oxidant stress* rather than a precise indicator of rates of $H_2O_2$ formation.

*2.2.3. Protocol 3: Estimation of Cellular $O_2^{\bullet-}$ Formation With Hydroethidine and FACS Analysis: Redox Cycling With Menadione*

Hydroethidine (HE) is cell permeable and reacts with $O_2^{\bullet-}$ to form ethidium (E), which in turn is believed to intercalate with DNA, providing nuclear fluorescence at an excitation wavelength of 520 nm and an emission wavelength of 610 nm *(26)*. Bindokas and coworkers reported that oxidation of HE by $O_2^{\bullet-}$ was specific because the reaction did not occur in the presence of other ROI, including $OH^{\bullet}$, $H_2O_2$, or $ONOO^-$ *(27)*. Because of apparent selectivity, HE has frequently been used to detect intracellular $O_2^{\bullet-}$ *(26)*.

2.2.3.1. Materials

1. HE (Molecular Probes).
2. PBS, pH 7.4 in 10 m$M$ glucose.
3. FACS machine or fluorescence plate reader (excitation wavelength 520 nm and emission wavelength 610 nm).

2.2.3.2. Method

1. Culture leukemic CEM suspension cells or HL60 suspension cells in RPMI with 10% FCS and supplements. Maintain cells in log-phase and keep them at a concentration between 2.5 and $5 \times 10^5$/mL.

2. After treatment (for example, with the redox cycling compound menadione, which generates $O_2^{\bullet-}$), centrifuge cells gently to remove culture media and wash twice by centrifugation in PBS (1,000 rpm).

3. Suspend cells in PBS, by light vortexing, and load with 5–10 $\mu M$ HE, incubating them for 15–20 min at 37°C.

4. Wash cells 2× in PBS (centrifuge for 1 min at 100$g$ and suspend in PBS containing 10 m$M$ glucose).

5. Because this cell assay is a viable one, immediate FACS analysis is required with the use of appropriate software for analysis.

6. The FACS setting is FL-3 (PE) in log mode after cell debris have been electronically gated out.

2.2.4. Protocol 4:

Estimation of Cellular $O_2^{\bullet-}$ Formation With HE and Fluorescence Microscopy: Induction by the Neutrophil Oxidant, Hypochlorous Acid

2.2.4.1. Materials

1. HE (Molecular Probes).

2. PBS, pH 7.4 in 10 m$M$ glucose.

3. Sodium hypochlorite solution.

4. Methionine (Sigma).

5. Fluorescence microscope with rhodamine filters capable of excitation wavelength 520 nm and emission wavelength 610 nm.

2.2.4.2. Method

1. Grow human hepatoma HepG2 cells in 24-well plates in minimal essential media and load with 5 $\mu M$ dihydroergotamine (DHE) or an equivalent volume of vehicle (DMSO) for 30 min. After this time wash cells with warm PBS (37°C) and then incubate for 5 min in PBS.

2. Determine the hypochlorite concentration at $\lambda$290 nm at pH 12 ($\varepsilon$ = 350/M/cm) and store on ice diluted in ice cold in ultra high-purity water (17). Immediately before use, dilute the hypochlorite solution in warm PBS and gently pour onto the HepG2 cells. To "quench" hypochlorite, 1 m$M$ methionine solution can be added for 1–2 min before the addition of hypochlorite.

3. Fluorescence is measured as a function of time or hypochlorite addition. Typical data obtained are shown in **Fig. 3.3**.

2.2.4.3. Comments

The intracellular oxidation of HE to E by $O_2^{\bullet-}$ has previously been analyzed with the use of flow cytometry and also by visualization of adherent neuronal cells and brain tissue with digital imaging microfluorometry (27). HE has also been used to study the respiratory burst in immune cells (26) and the redox state

**Untreated**          **HOCl**          **HOCl + methionine**

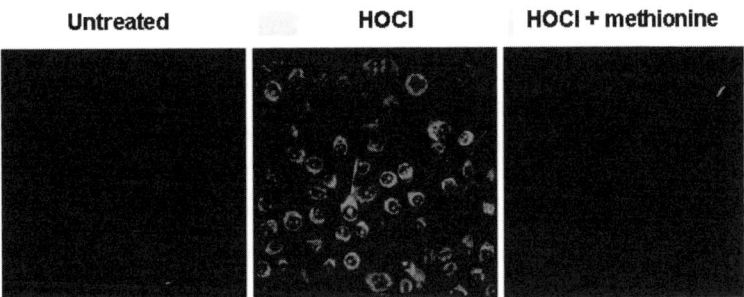

Fig. 3.3. Determination of intracellular 'ROI' produced by human hepatoma HepG2 cells exposed to the neutrophil oxidant, hypochlorite by fluorescence microscopy, and DHE.

in tumor cells *(28)*. To show the potential use of this probe, we have included an example of oxidation of HE to E by intracellular ROI generated by the neutrophil derived oxidant species, HOCl (**Fig. 3.3**). However, as with DCF, there are a number of problems associated with the use of HE as a quantitative marker of $O_2^{\bullet-}$ production *(22)*.

First, the amount of E produced by the oxidation of HE decreases with increasing $O_2^{\bullet-}$ flux, which suggests to us that HE catalyses the dismutation of $O_2^{\bullet-}$, which will underestimate the extent of $O_2^{\bullet-}$ production *(29)*. HE also can be oxidized by a variety of heme proteins, including mitochondrial cytochromes, hemoglobin, and myoglobin *(30)*. Most importantly, recent work has indicated that $O_2^{\bullet-}$ does not oxidize HE to E because it was found that $O_2^{\bullet-}$ generated by a variety of enzymatic and chemical systems (e.g., xanthine/xanthine oxidase, endothelial nitric oxide synthase, or potassium superoxide) oxidized HE to a fluorescent product (excitation, 480 nm; emission, 567 nm) that was different from E *(31)*. The authors concluded that the reaction between $O_2^{\bullet-}$ and HE formed a fluorescent marker product that was not E. Although not entirely specific for $O_2^{\bullet-}$, the method continues to be widely used for the "specific" determination of this radical species. We suggest that when HE is used for the determination of $O_2^{\bullet-}$ production, the combined use of pharmacological inhibitors of the mitochondrial ETC, together with the use of appropriate cell lines lacking mitochondrial DNA (i.e., without a functional ETC; generated as described in **ref**. *21)* be used to clarify the subcellular source of the $O_2^{\bullet-}$ signal.

***2.3. Chemiluminescence Procedures***

Luminol has been used as a detector of ROI formed during the reaction of XO with hypoxanthine (HX) for several decades *(32)*. Light produced by luminol + XO/HX is directly proportional to the activity of XO, suggesting that this system could be used as a general test for oxidizing intermediates *(32)*. Lucigenin (bis-*N*-methylacridinium) has also frequently been used for the luminescent detection of the radical $O_2^{\bullet-}$ by activated phagocytes or the XO/HX system *(33)*. More recently, several other compounds have been used for chemiluminescent detection of $O_2^{\bullet-}$, including coelenterazine [2-(4-hydroxybenzyl)-6-(4-hydroxyphenyl)-8-benzyl-3,7-dihydroimidazo [1,2-α]pyrazin-3-one] and its analogs CLA (2-methyl-6-phenyl-3,7-dihydroimidazo[1,2-α]pyrazin-3-one) and MCLA [2-methyl-6-(4-methoxyphenyl)-3,7-dihydroimidazo[1,2-α]pyrazin-3-one].

*2.3.1. Protocols: Chemiluminescence Characterization of ROI Generated by HX/XO and Phorbol Ester Stimulated White Blood Cells*

Luminol and lucigenin are often used to detect the production of ROI by activated phagocytes, although they have also been used to determine ROI in other cell types *(34)*.

2.3.1.1. Materials

1. 1 m*M* HX (Sigma-Aldrich, St. Louis, MO).
2. Copper and zinc-containing superoxide dismutase (CuZn-SOD; Sigma-Aldrich, St. Louis, MO).
3. DMSO (Sigma-Aldrich).
4. Luminol (Sigma-Aldrich).
5. XO (Sigma-Aldrich).
6. Ethylene diamine tetra-acetic acid (EDTA).
7. Catalase (CAT; Sigma-Aldrich).
8. PBS, pH 7.4.

2.3.1.2. Methods

A. **Luminol-dependent characterization of ROI generated by XO/HX system**

1. Add 100 µL of 1 m*M* HX in PBS to 50 µL of either CuZn-SOD (100 U/mL), CAT ($4 \times 10^3$ U/mL), or DMSO (100 m*M*) and add 100 µL of chemiluminescence (CL) probe (1 m*M* luminol) to the test wells of a white 96-well microtiter plate (in triplicate).
2. Start the reaction by the addition of 50 µL of XO (0.25 U/mL) using the automated luminometer dispenser.
3. Operate the luminometer in the integration mode for a time of 10 s with Ascent software (LabSystems, MN).
4. Express results as total integrated CL signal (relative light units; RLU).
5. These protocols may require substantial optimization to achieve maximum sensitivity and reliability of results.

B. **Luminol-dependent characterization of ROI generated by white blood cells**

1. Allow EDTA anticoagulated whole blood to sediment under gravity for 1 h at 37°C and remove the plasma and centrifuge

at ~400$g$ for 10 min to pellet out the white blood cells (WBCs).

2. Wash WBCs in PBS at 4°C and resuspend in PBS to a concentration of 2–5 × 10$^6$ cells/mL.

3. Add WBCs (100 µL) to a white 96-well microtitre plate in triplicate together with 50 µL of phorbol myristate acetate (PMA; 2 ng/mL) and 50 µL of test compounds including either CuZnSOD (100 U/mL), CAT (4 × 10$^3$ U/mL), or DMSO (100 m$M$).

4. Add 100 µL of 1 m$M$ luminol giving a final total volume of 300 µL.

5. Operate the luminometer in the kinetic mode at 10-s intervals for a total of 90 min using Ascent software.

6. Use the antioxidants SOD, CAT, and DMSO to characterize relative proportions of $O_2^{\bullet-}$, $H_2O_2$, and $OH^\bullet$ generated by PMA-stimulated WBCs.

7. Express the results either as the total integrated CL signal (RLU) or as RLUs per minute (RLU/min).

**2.3.1.3. Comments**

Although luminol and lucigenin are both used for the determination of $O_2^{\bullet-}$, the two reactions are different in that luminol requires univalent oxidation and lucigenin requires univalent reduction before they can react with $O_2^{\bullet-}$ and produce luminescence *per se*. In addition, they do not react with $O_2^{\bullet-}$ *(34, 35)*. A major problem with lucigenin is that the radical cation formed by its reduction can auto-oxidize and generate artifactual $O_2^{\bullet-}$. Because of this reaction, it has been suggested that lucigenin can be used for assaying SOD activity but it should not be used for measuring $O_2^{\bullet-}$ *(36)*. The extent to which this artifact can interfere with accurate measurement of $O_2^{\bullet-}$ by lucigenin is controversial *(7)*, but it appears to be significant in some cell systems *(35, 37–39)*. It has been suggested by some that by modulating the concentrations of lucigenin, it is possible to circumvent the problems associated with the overestimation of $O_2^{\bullet-}$. Because it is not possible to predict the extent of redox cycling, however, this may be difficult to accurately determine *(6)*. There are other problems associated with use of lucigenin. For example, the conversion of lucigenin to the radical by $O_2^{\bullet-}$ is not rapid and requires other cellular-reducing systems (e.g., XO, the mitochondrial electron transport chain, or the phagocyte NADPH oxidase), thus complicating data interpretation enormously *(35, 37, 38, 40)*. There are similar problems with the use of luminol for the detection of radicals such as $O_2^{\bullet-}$ because the radical intermediate formed on oxidation also auto-oxidizes *(6)*. Furthermore, luminol luminescence is also not a reliable indicator of $O_2^{\bullet-}$, even when $O_2^{\bullet-}$ is involved in the reaction leading to light emission, because it can mediate $O_2^{\bullet-}$ formation by a variety of oxidants including ferricyanide, OCl$^-$, and XO/HX *(41)*.

The problems of luminol and lucigenin associated with the artifactual generation of $O_2^{•-}$ can be circumvented by the use of more specific alternative nonredox-cycling compounds for the determination of $O_2^{•-}$ including coelenterazine, a luminophore isolated from the coelenterate Aequorea *(6, 7, 37, 40)*, and the Cypridina luciferin analogues CLA and MCLA, of which MCLA is the most sensitive probe for $O_2^{•-}$. Although the intensity of light emitted from the interaction of coelenterazine with $O_2^{•-}$ is greater than either that of lucigenin or luminol, coelenterazine-dependent chemiluminescence is not entirely specific for $O_2^{•-}$, because ONOO⁻ will also cause coelenterazine to luminescence *(37)*. A similar problem is found with MCLA which reacts with peroxyl radicals *(43)*. Our investigations with MCLA has shown that although it is more sensitive to $O_2^{•-}$ than luminol, it has a high level of background luminescence, is both light and temperature sensitive, and it auto-oxidizes, adding to the practical problems associated with nonspecificity. As such, it is recommended that all experiments are performed as quickly as possible in the dark and that reagents are similarly prepared in the dark.

***2.4. Electron Paramagnetic Resonance Spectroscopy***

The only technique that specifically detects free radicals is electron paramagnetic resonance (EPR) spectroscopy because it unequivocally measures the presence of unpaired electrons. However, unpaired electrons of species such as $O_2^{•-}$, •OH, and •NO are highly reactive radicals and generally do not accumulate to high-enough levels to be measured. One solution to the problem has been to use "spin traps" or "probes" that intercept reactive radical intermediates and form stable longer-lasting radical adducts with characteristic EPR signatures *(1)*. These adducts accumulate to levels that can be detected with the use of EPR spectroscopy and provide information to enable the identification of the originating free radical species. A wide range of spin traps are commercially available for use both in whole animal studies and cell culture systems, including *N*-tertbutyl-*p*-phenylnitrone and 5,5-dimethyl-1-pyrroline *N*-oxide (DMPO) *(44)*. Recently, newer compounds including 1,1,3-trimethyl-isoindole *N*-oxide *(45)*, *N*-2-(2-ethoxycarbonyl-propyl)-a-phenylnitrone *(46)* and 5-diethoxyphosphoryl-5-methyl-1-pyrroline *N*-oxide (DEPMPO) reported to be more $O_2^{•-}$ specific and stable have also emerged *(44, 47, 48)*.

***2.4.1. Protocol: EPR Spectroscopy Characterization of ROI Generated by Xanthine Oxidase Using the Spin-Trap DEPMPO***

***2.4.1.1. Materials***

1. Chelex-100.
2. PBS, pH 7.4.
3. DEPMPO (Oxis International Inc., Portland, OR).
4. HX (Sigma-Aldrich).
5. XO (Sigma-Aldrich).
6. Diethylene triamine pentaacetic acid (Sigma-Aldrich).

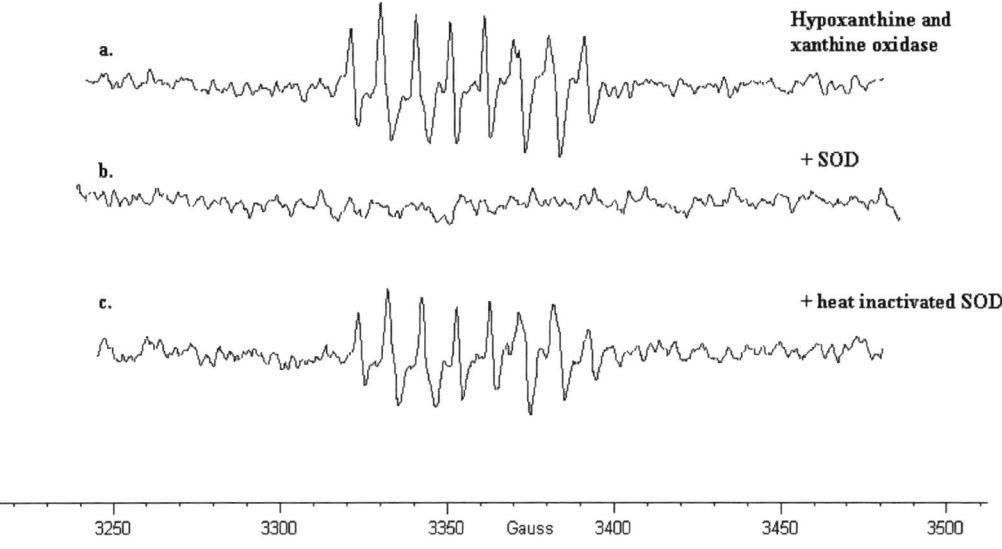

a. 5 µM hypoxathine, 10 mM DEPMPO, 0.1 mM DTPA and 0.01 U/ml xanthine oxidase

b. 5 µM hypoxathine, 10 mM DEPMPO, 0.1 mM DTPA, 400 U/ml SOD and 0.01 U/ml xanthine oxidase

c. 5 µM hypoxathine, 10 mM DEPMPO, 0.1 mM DTPA, 400 U/ml SOD and 0.01 U/ml xanthine oxidase

Centre field – 3362

Fig. 3.4. Determination by electron paramagnetic resonance of $O_2^{\bullet-}$ generated by xant1hine/xanthine oxidase using the spin-trap DEPMPO (see text for details). DEPMPO- $O_2^{\bullet-}$ adduct is inhibited by the addition of SOD but not by heat inactivated SOD.

**2.4.1.2. Method**

1. Pretreat PBS buffer and water with the ion-exchange resin Chelex-100 (1 g/100 mL; Bio-Rad, Hercules, CA) to remove adventitious metals in the buffer (49) and add diethylene triamine pentaacetic acid at a final concentration of 0.1 m$M$.

2. Incubate hypoxanthine with xanthine oxidase (0.01 units/mL) in the presence of the spin trap DEPMPO (10 m$M$).

3. Add the mixtures into an aqueous quartz flat cell (Wilmad, Buena, NJ), which should then be centred in a TE011 cavity.

4. Record EPR spectra with a Bruker 200D spectrometer operated at 9.7 GHz with a 1-GHz modulation frequency.

5. Transfer the data to a computer for simulation analysis (49). Typical data obtained under these conditions are illustrated in **Fig. 3.4**.

**2.4.1.3. Comments**

The determination of radical species such as $O_2^{\bullet-}$, $^{\bullet}OH$, and $^{\bullet}NO$ by EPR spectroscopy requires the use of spin-traps to generate species with longer half-lives required for their detection by EPR spectroscopy. The technique can also be used to detect more stable free radical-derived species, including ascorbyl radical, tocopheroxyl radical, and heme–nitrosyl complexes produced in vascular tissues during oxidative injury and inflammation *(50)*. However, there are a several potential problems with the use of spin traps that require consideration *(1, 44, 51)*. For example, one must consider whether the reaction products give the EPR signal can be rapidly removed in vivo and in cultured cells by enzymatic metabolism and also by direct reduction by agents such as ascorbate. For example, when DMPO is used to trap the $^{\bullet}OH$, ascorbate can directly reduce the DMPO-hydroxyl radical adduct to an EPR-silent species *(1)*. DMPO can also be oxidized by ferric ions, which generates the four-line spectrum normally produced by $^{\bullet}OH$ *(52)*. Several published studies reporting on the use of spin traps to detect radical species frequently fail to show use of the proper controls *(53)*. For example, does the added compound interfere with the radical-generating system (e.g., decomposing $H_2O_2$ or chelating iron in the $Fe^{2+}/H_2O_2$ system), or does it interact directly with the trap spin adduct, reducing it to an EPR-silent species?

Rizzi et al. have introduced triarylmethyl free radical, TAM OX063, as a probe for the detection of $O_2^{\bullet-}$ in aqueous solution *(54)*. In this case, the $O_2^{\bullet-}$ reacts with the probe to cause the loss of the EPR signal. One advantage of TAM OX063 is that it is not subject to reduction by such agents as ascorbate or reduced thiols such as glutathione (GSH). Valgimigli and colleagues described an EPR method for the measurement of the oxidative stress status in biological systems *(55)*. The method was based on the X-band EPR detection of a nitroxide generated under physiological conditions by oxidation of bis(1-hydroxy-2,2,6,6-tetramethyl-4-piperidinyl)-decandioate which is administrated as hydrochloride salt. Since the probe is reported to react rapidly with the majority of radical species involved in the oxidative stress and cross cell membranes easily it was suggested to be applicable in the clinical setting *(55)*.

# 3. Measurement of ROI in the Mitochondria

### 3.1. Isolation of Mitochondria From Cells and Tissue

#### 3.1.1. Animal Cells

1. Pellet approximately $2 \times 10^7$ cells by centrifugation in a microcentrifuge tube at $1000g$ for 3 min.

2. Carefully remove and discard the supernatant.

3. Wash cells in MSHE buffer (sucrose 70 m$M$, mannitol 220 m$M$, HEPES 2 m$M$, ethylene glycol tetraacetic acid (EGTA), bovine serum albumin (BSA) 0.5 m$M$ @ pH 7.4) and centrifuge at $1000g$ for 3 min.

4. Remove supernatant and suspend cells in approximately ~150 µL of fresh MSHE buffer.

[Au7]

5. Syringe homogenize cells through a needle (27.5 gage) until 80–90% cells are blue by trypan blue exclusion.

6. Centrifuge cells at $1000g$ for 10 min at 4°C and transfer the supernatant to a new tube, which is further centrifuged at $10,000g$ for 20 min at 4°C.

7. Transfer the supernatant from the $10,000g$ centrifugation to a new microcentrifuge tube -this contains the cytosolic fraction. The pellet from the $10,000g$ centrifugation contains the isolated mitochondrial fraction. Maintain the mitochondria pellet in MSHE buffer on ice prior to processing. Mitochondrial fractions should support coupled respiration analysed using a Clark-type electrode or equivalent.

*3.1.2. Animal Tissue: Isolation of Rat Liver Mitochondria*

We have found liver preparations from laboratory rodents to be suitable for large quantities of good quality and coupled mitochondria *(16, 17)* obtained using the following protocol.

1. Use male Sprague–Dawley rats weighing 180–220 g for isolation of liver mitochondria. Rat liver mitochondria were isolated as follows.

2. Chop fresh liver tissues finely and homogenize in ice-cold isolation medium containing 220 m$M$ mannitol, 70 m$M$ sucrose, 2 m$M$ HEPES, 0.5 m$M$ EGTA, and 0.1% BSA (fat free) (pH 7.4), using a Dounce homogenizer.

3. Centrifuge the homogenate at $1000g$ for 10 min at 4°C in a Beckman JA20 rotor.

4. Transfer the supernatant into a fresh tube and centrifuge as described previously.

5. Centrifuge the supernatant at $10,000g$ for 10 min at 4°C and retain the pellet.

6. Scrap the middle dark brown layer from the pellet and transfer to another fresh tube.

7. Suspend the final pellet in a small volume of the isolation buffer.

8. This sample may now be assayed for protein concentration.

9. The condition of the intact mitochondria can be tested by measuring oxygen consumption in the presence of succinate and ADP and determining respiratory control ratio *(21)*.

***3.2. Estimation of Mitochondrial 'ROI' (H₂O₂)***

In the presence of horseradish peroxidase (HRP), Amplex red reagent (10-acetyl-3,7-dihydroxyphenoxazine) reacts with $H_2O_2$ with a 1:1 stoichiometry to produce a highly fluorescent resorufin product. Amplex red/HRP has been observed to detect levels of $H_2O_2$ produced by the NADPH oxidase and other oxidases *(56)*.

*3.2.1. Materials*

1. Amplex red (Molecular Probes).

2. HRP (Molecular Probes).

3. Krebs-Ringer phosphate (KRP: 145 m$M$ NaCl, 5.7 m$M$ sodium phosphate, 4.86 m$M$ KCl, 0.54 m$M$ CaCl$_2$, 1.22 m$M$ MgSO$_4$, 5.5 m$M$ glucose, pH 7.35).

4. Fluorescence plate reader such as the SpectraMax Plus spectrophotometer (Molecular Devices Corporation, Sunnyvale, CA; excitation wavelength 560 nm and emission wavelength 590 nm).

5. Standard curves are constructed by adding known amounts of H$_2$O$_2$ to assay medium in the presence of the reactants (amplex red and HRP).

*3.2.2. Method*

1. Culture human Jurkatt suspension cells in RPMI 1640 with 10% FCS and supplements and maintain in log-phase. Keep at a concentration between 2.5 and 5 × 10$^5$/mL.

2. Suspend the mitochondrial pellet obtained from approximately 40 mL of 1.5 × 10$^6$ cells in 200 µL of KRP buffer to form the mitochondrial mixture. Keep on ice before use.

3. Add 100 µL of HRP/amplex red (0.1 unit/mL and 50 µ$M$, respectively) in KRP buffer to each microplate well and initiate the reaction by adding 20 µL of the mitochondrial mixture.

4. For negative control, add 20 µl of KRP alone.

5. Measure fluorescence at 15-s intervals for a total time of 30 min. The initial linear phase of increase in absorbance is used to calculate the rate of hydrogen peroxide production per milligram of mitochondrial protein. Background fluorescence is then measured in the absence of mitochondria and results are calculated as RFU mitochondrial sample minus background RFU and expressed as pmol H$_2$O$_2$/mg of protein/30 min).

6. Construct a standard curve by adding a known amount of H$_2$O$_2$ to the assay medium in the absence of mitochondria (linearity up to approx 2 µM). Under the aforementioned assay conditions, SOD does not increase the rate of production of H$_2$O$_2$, indicating that the nonenzymatic dismutation of O$_2^{\bullet-}$ to H$_2$O$_2$ is rapid and complete.

**3.3. Measurement of Mitochondrial ROI (O$_2^{\bullet-}$): The Determination of Mitochondrial ROI With DHE and Gemini XS Dual Monochromatic Fluorescence Plate Reader**

DHE can be used to estimate mitochondrial 'ROI' formation, in this case O$_2^{\bullet-}$, by isolated mitochondria.

*3.3.1. Materials*

1. MSE buffer: 220 m$M$ mannitol, 70 m$M$ sucrose, 2 m$M$ HEPES, 0.5 m$M$ EGTA, 0.1% BSA (fat free) (pH 7.4).

2. Rotenone (2 µ$M$; Sigma-Aldrich).

3. Succinate (10 m$M$; Sigma-Aldrich).

4. Antimycin A (10 µ$M$; Sigma-Aldrich).

5. ADP (250 µ$M$; Sigma-Aldrich).

6. KH$_2$PO$_4$ (5 m$M$).

7. Fluorescence is measured in black 96-well microplates on a Gemini XS dual monochromatic reader at an excitation/ emission wavelength of 520/610 nm.

*3.3.2. Method*

1. For the determination of mitochondrial ROI under state 4 (resting) conditions, reagents should be added in the following sequence (total volume 200 µL): MSE buffer, mitochondria (40 µg), succinate (10 m$M$)/rotenone (2 µ$M$) mix, and DHE (1 µ$M$).

2. For experiments with mitochondria under state 3 (ADP-stimulated) conditions, add ADP (250 µ$M$) and KH$_2$PO$_4$ (5 m$M$) after the succinate/rotenone mix.

3. Conduct the assay in triplicate, and follow the fluorescence kinetically over 15 min (at 37°C) in a Gemini XS dual monochromatic reader at an excitation/emission wavelength of 520/610 nm, respectively.

4. For mitochondria, use antimycin A (10 µ$M$) as a positive control.

# 4. General Problems Associated With the Measurement of ROI

We have briefly described several specific methodological problems and caveats associated with measuring ROI formation in isolated cells and mitochondria. There are also a number of general problems that should be taken into account. For example, many methods that allegedly can be used to measure ROI have become widely used, but precise information on what really is measured with these methods is conspicuously lacking. An important first consideration is that cell culture itself induces oxidative stress, both by facilitating generation of ROI and by hindering the adaptive upregulation of cellular antioxidants (reviewed in **ref**. *57*). It has even been suggested that the "Hayflick limit" in fibroblasts (replicative senescence after a certain number of cell divisions) may be an artifact of oxidative stress imposed during cell culture *(58)*.

In addition, results can be confounded by free radical reactions taking place in the culture media itself *(25, 59–61)*. For example, a number of reports of effects of ascorbate and polyphenolic compounds (e.g., flavonoids) on cells in culture appear to be largely artifactual and caused by the oxidation of these compounds in the culture media *(62, 63)*. Therefore, it is important to consider what reactions take place in the cell culture medium alone when compounds are added to it (e.g., **ref**. *62, 64)*. Furthermore, the presence of cells can also suppress free radical reactions occurring in the medium *(57)*. Some simple principles can be used as guidelines in understanding oxidative stress/oxidative damage in cell culture. Hydrogen peroxide generally crosses cell membranes readily, probably *via* aquaporins *(65)*. Thus, catalase added outside cells can exert both intracellular and extracellular effects on $H_2O_2$ level, the former by "draining" $H_2O_2$ out of the cell by removing extracellular $H_2O_2$ and thus establishing a concentration gradient *(57)*. In contrast, $O_2^{\bullet-}$ does not readily cross cell membranes readily *(66, 67)*.

Therefore, if the addition of external SOD is protective against an event in cell culture, one should be cautious as to what these data really mean, as they could be indicative of extracellular $O_2^{\bullet-}$ generating reactions. To further complicate this issue, if the addition of extracellular SOD is damaging to cells in culture, this could suggest that an extracellular $O_2^{\bullet-}$-generating reaction is still occurring because the SOD can increase levels of $H_2O_2$, which will rapidly enter the cell and inhibit cellular ATP production *(49)*. Similarly, neither the iron-chelating agent deferoxamine (which suppresses most, but not all, iron-dependent free radical reactions) nor the thiol antioxidant GSH readily enter cells. Therefore, caution is required again if these agents elicit short-term protective effects, because this would suggest extracellular oxidant production *(67–69)*. Recently, Clement et al. *(64)* showed that GSH protected against the cytotoxicity of dopamine simply because it reacted with dopamine oxidation products generated in the cell culture medium. However, given long enough, more or less everything can enter cells, including deferoxamine, superoxide dismutase (SOD) and GSH *(70, 71)*.

Studies of ROI production by mitochondria is an area of particular interest but also of great difficulty too. DHE is often use to determine mitochondrial production of $O_2^{\bullet-}$ in situ but since it is clear that DHE responds to extra-mitochondrially produced $O_2^{\bullet-}$ by redox cycling agents such as menadione or paraquat, the use of DHE to specifically indicate mitochondrial $O_2^{\bullet-}$ production may often be erroneous. Increased specificity can be attained by the use of pharmacological inhibitors of mitochondrial electron transport provided one has performed appropriate control experiments and is confident that the compounds used to do not "scavenge" ROI or interfere with their measurement. Additional controls using genetically modified cells lacking mitochondrial DNA (mtDNA) to determine whether or not the $O_2^{\bullet-}$ is mitochondrial or extramitochondrial origin may increase the specificity of data analysis *(21)*. A simpler method, although one not available to most laboratories is the use of aromatic compounds (such as salicylate or phenylalanine) or spin traps as described above. Spin traps have been used successfully in many cell studies, because a wider range of traps at higher concentrations can be used than could ever be employed in vivo. Again, one must be aware of the possibility of rapid reduction of free radical–spin trap adducts to EPR-silent species by nonenzymatic antioxidants (such as ascorbate) and cellular enzymatic reducing systems. An interesting combination is 5-((2-carboxy)phenyl)-5-hydroxy-1-(2,2,5,5-tetramethyl-1-oxypyrrolidin-3-yl)methyl-3-phenyl-2-pyrrolin-4-one sodium salt, a nitroxide that is nonfluorescent. When it combines with a ROI, the nitroxide is removed, the EPR signal is lost and the fluorescence is restored *(72)*.

## 5. Conclusions

Whatever method is used to measure ROI, it is necessary to consider the methodological limitations and problems of specificity. Where appropriate, it is advisable to use more than one approach to determine the precise nature of the ROI being studied. Therefore, investigators are urged to think carefully about the methods mechanism, potential problems with specificity, and whether the method is quantitative. With appropriate care and attention as well as the judicious use of pharmacological inhibitors and genetically modified cells, erroneous interpretations can be minimised and the exciting field of free radical research can continue to move forward.

### References

1. Halliwell, B., and Gutteridge, J. M. C. (2007) *Free Radicals in Biology and Medicine* (4th Ed). Oxford University Press, Oxford, England. ISBN-10: 019856869X

2. Lambert, A. J. and Brand, M. D. (2004) Inhibitors of the quinone-binding site allow rapid superoxide production from mitochondrial NADH:ubiquinone oxidoreductase (complex I). *J Biol Chem* **279**, 39414–394120.

3. Turrens, J. F. (1997) Superoxide production by the mitochondrial respiratory chain. *Biosci Rep* **17**, 3–8.

4. Raha, S. and Robinson, B. H. (2000) Mitochondria, oxygen free radicals, disease and ageing. *Trends Biochem Sci* **25**, 502–508.

5. Li, Y., Huang, T. T., Carlson, E. J., Melov, S., Ursell, P. C., Olson, J. L., Noble, L. J., Yoshimura, M. P., Berger, C., Chan, P. H., Wallace, D. C., and Epstein, C. J. (1995) Dilated cardiomyopathy and neonatal lethality in mutant mice lacking manganese superoxide dismutase. *Nat Genet* **11**, 376–381.

6. Tarpey, M. M. and Fridovich, I. (2001) Methods of detection of vascular reactive species: nitric oxide, superoxide, hydrogen peroxide, and peroxynitrite. *Circ Res* **89**, 224–236.

7. Munzel, T., Afanas'ev, I. B., Kleschyov, A. L., and Harrison, D. G. (2002) Detection of superoxide in vascular tissue. *Arterioscler Thromb Vasc Biol* **22**, 1761–1768.

8. Auclair, C. and Voisin, E. (1985) Nitroblue tetrazolium reduction, in *CRC Handbook of Methods for Oxygen Radical Research* (Greenwald R. A., ed.), CRC Press, Boca Raton, FL, pp. 123–132.

9. Grozdanovic, Z., Nakos, G., Christova, T., Nikolova, Z., Mayer, B., and Gossrau, R. (1995) Demonstration of nitric oxide synthase (NOS) in marmosets by NADPH diaphorase (NADPH-d) histochemistry and NOS immunoreactivity. *Acta Histochem* **97**, 321–331.

10. Thomson, L., Trujillo, M., Telleri, R., and Radi, R. (1995) Kinetics of cytochrome c oxidation by peroxynitrite: implications for superoxide measurements in nitric oxide-producing biological systems. *Arch Biochem Biophys* **319**, 491–497.

11. Azzi, A., Montecucco, C., and Richter, C. (1975) The use of acetylated ferricytochrome c for the detection of superoxide radicals produced in biological membranes. *Biochem Biophys Res Commun* **65**, 597–603.

12. Kuthan, H., Ullrich, V., and Estabrook, R. W. (1982) A quantitative test for superoxide radicals produced in biological systems. *Biochem J* **203**, 551–558.

13. Gardner, P. R. and Fridovich, I. (1991) Superoxide sensitivity of the *Escherichia coli* aconitase. *J Biol Chem* **266**, 19328–19333.

14. Hausladen, A. and Fridovich, I. (1994) Superoxide and peroxynitrite inactivate aconitases, but nitric oxide does not. *J Biol Chem* **269**, 29405–29408.

15. Bass, D. A., Parce, J. W., Dechatelet, L. R., Szejda, P., Seeds, M. C., and Thomas, M. (1983) Flow cytometric studies of oxidative product formation by neutrophils: a graded response to membrane stimulation. *J Immunol* **130**, 1910–1917.

16. Whiteman, M., Armstrong, J. S., Jones, D. P., and Halliwell, B. (2004) Peroxynitrite mediates calcium-dependent mitochondrial dysfunction and cell death via activation of calpains. *FASEB J* **18**, 1395–1397.

17. Whiteman, M., Rose, P., Siau, J. L., Cheung, N. S., Tan, G. S., Halliwell, B., and Armstrong, J. S. (2005) Hypochlorous acid-mediated mitochondrial dysfunction and apoptosis in human hepatoma HepG2 and human fetal liver cells: role of mitochondrial permeability transition. *Free Rad Biol Med* **38**, 1571–1584.

18. Armstrong, J. S. and Jones, D. P. (2002) Glutathione depletion enforces mitochondrial permeability transition and apoptosis in HL60 cells overexpressing Bcl-2. *FASEB J* **16**, 1263–1265.

19. Armstrong, J. S., Whiteman, M., Yang, H., Jones, D. P., and Sternberg, P. (2004) Cysteine-starvation activates the redox-dependent mitochondrial permeability transition in retinal pigment epithelial cells. *IOVS* **45**, 4183–4189.

20. Armstrong, J. S., Yang, H., Duan, W., Chua, Y., and Whiteman, M. (2004) Cytochrome $bc_1$ regulates the mitochondrial permeability transition by two distinct pathways. *J Biol Chem* **279**, 50420–50428.

21. Whiteman, M., Chua, Y. L., Zhang, D., Duan, W., Liou, Y. C., and Armstrong, J. S. (2006) Nitric oxide blocks glutathione-dependent cell death independently of mitochondrial reactive oxygen species: potential role of s-nitrosylation? *Biochem Biophys Res Commun* **339**, 255–262.

22. Halliwell, B. and Whiteman, M. (2004) Measuring reactive species and oxidative damage *in vivo* and cell culture. How should you do it and what does it mean? *Br J Pharmacol* **142**, 231–255.

23. LeBel, C. P., Ischiropoulos, H., and Bondy, S. C. (1992) Evaluation of the probe 2′,7′-dichlorofluorescin as an indicator of reactive oxygen species formation and oxidative stress. *Chem Res Toxicol* **5**, 227–231.

24. Rota, C., Chignell, C. F., and Mason, R. P. (1999) Evidence for free radical formation during the oxidation of 2′-7′-dichlorofluorescin to the fluorescent dye 2′-7′-dichlorofluorescein by horseradish peroxidase: possible implications for oxidative stress measurements. *Free Radic Biol Med* **27**, 873–881.

25. Halliwell, B. and Whiteman, M. (2004) Measuring reactive species and oxidative damage *in vivo* and in cell culture: how should you do it and what do the results mean? *Br J Pharmacol* **142**, 231–255.

26. Rothe, G. and Valet, G. (1990) Flow cytometric analysis of respiratory burst activity in phagocytes with hydroethidine and 2,7-dichlorofluorescin. *J Leukocyte Biol* **47**, 440–448.

27. Bindokas, V. P., Jordan, J., Lee, C. C., and Miller, R. J. (1996) Superoxide production in rat hippocampal neurons: selective imaging with hydroethidine *J Neurosci* **16**, 1324–1326.

28. Olive, P. L. (1989) Hydroethidine: a fluorescent redox probe for locating hypoxic cells in spheroids and murine tumours. *Br J Cancer* **160**, 332–328.

29. Benov, L., Sztejnberg, L., and Fridovich, I. (1998) Critical evaluation of the use of hydroethidine as a measure of superoxide anion radical. *Free Radic Biol Med* **25**, 826–831.

30. Papapostolou, I., Patsoukis, N., and Georgiou, C. D. (2004) The fluorescence detection of superoxide radical using hydroethidine could be complicated by the presence of heme proteins. *Anal Biochem* **332**, 290–298.

31. Zhao, H., Kalivendi, S., Zhang, H., Joseph, J., Nithipatikom, K., Vasquez-Vivar, J., and Kalyanaraman, B. (2003) Superoxide reacts with hydroethidine but forms a fluorescent product that is distinctly different from ethidium: potential implications in intracellular fluorescence detection of superoxide. *Free Radic Biol Med* **34**, 1359–1368.

32. Totter, J. R., de Dugros, E. C., and Riveiro, C. (1960) The use of chemiluminescent compounds as possible indicators of radical production during xanthine oxidase action. *J Biol Chem* **235**, 1839–18342.

33. Storch, J. and Ferber, E. (1988) Detergent-amplified chemiluminescence of lucigenin for determination of superoxide anion production by NADPH oxidase and xanthine oxidase. *Anal Biochem* **169**, 262–267.

34. Faulkner, K. and Fridovich, I. (1993) Luminol and lucigenin as detectors for $O_2-$. *Free Radic Biol Med* **15**, 447–451.

35. Spasojevic, I., Liochev, S. I., and Fridovich, I. (2000) Lucigenin: redox potential in aqueous media and redox cycling with $O_2-$ production. *Arch Biochem Biophys* **373**, 447–450.

36. Liochev, S. I. and Fridovich, I. (1997) Lucigenin (bis-*N*-methylacridinium) as a mediator of superoxide anion production. *Arch Biochem Biophys* **337**, 115–120.

37. Tarpey, M. M., White, C. R., Suarez, E., Richardson, G., Radi, R., and Freeman, B. A. (1999) Chemiluminescent detection of oxidants in vascular tissue. Lucigenin but not coelenterazine enhances superoxide formation. *Circ Res* **84**, 1203–1211.

38. Sohn, H. Y., Keller, M., Gloe, T., Crause, P., and Pohl, U. (2000) Pitfalls of using lucigenin in endothelial cells: implications for NAD(P)H dependent superoxide formation. *Free Radic Res* **32**, 265–272.

39. Wardman, P., Burkitt, M. J., Patel, K. B., Lawrence, A., Jones, C. M., Everett, S. A., and Vojnovic, B. (2002) Pitfalls in the use of common luminescent probes for oxidative and nitrosative stress. *J Fluorescence* **12**, 65–68.

40. Tarpey, M. M., White, C. R., Suarez, E., Richardson, G., Radi, R., and Freeman, B. A. (1999) Chemiluminescent detection of oxidants in vascular tissue. Lucigenin but not coelenterazine enhances superoxide formation. *Circ Res* **84**, 1203–1211.

41. Hodgson, E. K., and Fridovich, I. (1973) The role of $O_2-$ in the chemiluminescence of luminol. *Photochem Photobiol* **18**, 451–455.

42. Teranishi, K. and Shimomura, O. (1997) Coelenterazine analogs as chemiluminescent probe for superoxide anion. *Anal Biochem* **249**, 37–43.

43. Tampo, Y., Tsukamoto, M., and Yonaha, M. (1998) The antioxidant action of 2-methyl-6-(p-methoxyphenyl)-3,7-dihydroimidazo[1,2-alpha]pyra z in-3-one (MCLA), a chemiluminescence probe to detect superoxide anions. *FEBS Lett* **430**, 348–352.

44. Khan, N., Wilmot, C. M., Rosen, G. M., Demidenko, E., Sun, J., Joseph, J., O'Hara, J., Kalyanaraman, B., and Swartz, H. M. (2003) Spin traps: *in vitro* toxicity and stability of radical adducts. *Free Radic Biol Med* **34**, 1473–1481.

45. Bottle, S. E., Hanson, G. R., and Micallef, A. S. (2003) Application of the new EPR spin trap 1,1,3-trimethylisoindole *N*-oxide (TMINO) in trapping HO. and related biologically important radicals. *Org Biomol Chem* **1**, 2585–2589.

46. Stolze, K., Udilova, N., Rosenau, T., Hofinger, A., and Nohl, H. (2003) Spin trapping of superoxide, alkyl- and lipid-derived radicals with derivatives of the spin trap EPPN. *Biochem Pharmacol* **66**, 1717–1726.

47. Frejaville, C., Karoui, H., Tuccio, B., Le Moigne, F., Culcasi, M., Pietri, S., Lauricella, R., and Tordo, P. (1995) 5-(Diethoxyphosphoryl)-5-methyl-1-pyrroline *N*-oxide: a new efficient phosphorylated nitrone for the *in vitro* and *in vivo* spin trapping of oxygen-centered radicals. *J Med Chem* **38**, 258–265.

48. Liu, K. J., Miyake, M., Panz, T., and Swartz, H. (1999) Evaluation of DEPMPO as a spin trapping agent in biological systems. *Free Radic Biol Med* **26**, 714–721.

49. Armstrong, J. S., Rajasekaran, M., Chamulitrat, W., Gatti, P. J., Hellstrom, W. J., and Sikka, S. C. (1999) The effects of reactive oxygen intermediates on human spermatozoa movement and energy metabolism. *Free Radic Biol Med* **26**, 869–880.

50. Laurindo, F. R., Pedro Mde, A., Barbeiro, H. V., Pileggi, F., Carvalho, M. H., Augusto, O., and da Luz, P. L. (1994) Vascular free radical release: ex vivo and in vivo evidence for a flow-dependent endothelial mechanism. *Circ Res* **74**, 700–709.

51. Rosen, G. M., Britigan, B., Halpern, H., and Pou, S. (1999) *Free Radicals Biology and Detection by Spin Trapping*, Oxford University Press, Oxford.

52. Makino, K., Hagiwara, T., Hagi, A., Nishi, M., and Murakami A. (1990) Cautionary note for DMPO spin trapping in the presence of iron ion. *Biochem Biophys Res Commun* **172**, 1073–1080.

53. Halliwell, B. (1995) Antioxidant characterization. Methodology and mechanism. *Biochem Pharmacol* **49**, 1341–1348.

54. Rizzi, C., Samouilov, A., Kutala, V. K., Parinandi, N. L., Zweier, J. L., and Kuppusamy, P. (2003) Application of a trityl-based radical probe for measuring superoxide. *Free Radic Biol Med* **35**, 1608–1618.

55. Valgimigli, L., Pedulli, G. F., and Paolini, M. (2001) Measurement of oxidative stress by EPR radical-probe technique. *Free Radic Biol Med* **31**, 708–716.

56. Zhou, M., Diwu, Z., Panchuk-Voloshina, N., and Haugland, R. P. (1997) A stable nonfluorescent derivative of resorufin for the fluorometric determination of trace hydrogen peroxide: Applications in detecting the activity of phagocyte NADPH oxidase and other oxidases. *Anal Biochem* **253**, 162–168.

57. Halliwell, B. (2003) Oxidative stress in cell culture: an under-appreciated problem? *FEBS Lett* **540**, 3–6.

58. Wright, W. E. and Shay, J. W. (2002) Historical claims and current interpretations of replicative aging. *Nat Biotechnol* **20**, 682–688.

59. Grzelak, A., Rychlik, B., and Bartosz, G. (2000) Reactive oxygen species are formed in cell culture media. *Acta Biochim Pol* **47**, 1197–1198.

60. Roques, S. C., Landrault, N., Teissedre, P. L., Laurent, C., Besancon, P., Rouane, J. M., and Caporiccio, B. (2002) Hydrogen peroxide generation in caco-2 cell culture medium by addition of phenolic compounds: effect of ascorbic acid. *Free Radic Res* **36**, 593–599.

61. Wee, L. M., Long, L. H., Whiteman, M., and Halliwell, B. (2003) Factors affecting the ascorbate- and phenolic-dependent generation of hydrogen peroxide in Dulbecco's Modified Eagles Medium. *Free Radic Res* **37**, 1123–1130.

62. Clement, M. V., Ramalingam, J., Long, L. H., and Halliwell, B. (2001) The *in vitro* cytotoxicity of ascorbate depends on the culture medium used to perform the assay and involves hydrogen peroxide. *Antioxid Redox Signal* **3**, 157–163.

63. Long, L. H. and Halliwell, B. (2001) Antioxidant and prooxidant abilities of foods and beverages. *Methods Enzymol* **335**, 181–190.

64. Clement, M. V., Long, L. H., Ramalingam, J., and Halliwell, B. (2002) The cytotoxicity of dopamine may be an artifact of cell culture. *J Neurochem* **81**, 414–421.

65. Henzler, T. and Steudle, E. (2000) Transport and metabolic degradation of hydrogen peroxide in Chara corallina: model calculations and measurements with the pressure probe suggest transport of $H_2O_2$ across water channels. *J Exp Bot* **51**, 2053–2066.

66. Lynch, R. E. and Fridovich, I. (1978) Permeation of the erythrocyte stroma by superoxide radical. *J Biol Chem* **253**, 4697–4699.

67. Marla, S. S., Lee, J., and Groves, J. T. (1997) Peroxynitrite rapidly permeates phospholipid membranes. *Proc Natl Acad Sci USA* **94**, 14243–14248.

68. Meister, A. and Anderson, M. E. (1983) Glutathione. *Ann Rev Biochem* **52**, 711–760.

69. Halliwell, B. (1989) Protection against tissue damage in vivo by desferrioxamine: what is its mechanism of action? *Free Radic Biol Med* **7**, 645–651.

70. Doulias, P. T., Christoforidis, S., Brunk, U. T., and Galaris, D. (2003) Endosomal and lysosomal effects of desferrioxamine: protection of HeLa cells from hydrogen peroxide-induced DNA damage and induction of cell-cycle arrest. *Free Radic Biol Med* **35**, 719–728.

71. Rius, M., Nies, A. T., Hummel-Eisenbeiss, J., Jedlitschky, G., and Keppler, D. (2003) Cotransport of reduced glutathione with bile salts by MRP4 (ABCC4) localized to the basolateral hepatocyte membrane. *Hepatology* **38**, 374–384.

72. Pou, S., Huang, Y. I., Bhan, A., Bhadti, V. S., Hosmane, R. S., Wu, S. Y., Cao, G. L., and Rosen, G. M. (1993) A fluorophore-containing nitroxide as a probe to detect superoxide and hydroxyl radical generated by stimulated neutrophils. *Anal Biochem* **212**, 85–90.

# Chapter 4

# Redox-Sensitive Green Fluorescent Protein: Probes for Dynamic Intracellular Redox Responses. A Review

## Mark B. Cannon, and S. James Remington

## Abstract

The quantification of transient redox events within subcellular compartments, such as those involved in certain signal transduction pathways, requires specific probes with high spatial and temporal resolution. Redox-sensitive variants of the green fluorescent protein (roGFP) have recently been developed that allow the noninvasive monitoring of intracellular thiol-disulfide equilibria. In this chapter, the biophysical properties of these probes are discussed, including recent efforts to enhance their response times. Several recent applications of roGFPs are highlighted, including roGFP expression within *Arabidopsis* to monitor redox status during root elongation, expression in neurons to measure oxidative stress during ischemia, and targeting of roGFPs to endosomal compartments demonstrating unexpectedly oxidizing potentials within these compartments. Possible future directions for the optimization of roGFPs or new classes of redox-sensitive fluorescent probes are also discussed.

**Keywords:** Biosensor, disulfide, GFP, midpoint potential, protein engineering, redox.

## 1. Introduction

Cysteine residues in proteins are reactive and play many key roles in enzymatic activity as well as in protein folding, for example, by stabilizing folded states through intramolecular disulfide bond formation. More recently, it has come to light that cysteine reactivity, in particular toward reactive oxygen and nitrogen species (ROS and RNS), is also critical for dynamic processes such as gene regulation, oxidative stress response, and cell signaling *(1)*. Despite the increasingly clear significance and apparent complexity of thiol chemistry in living systems, the conventional method

J.T. Hancock (ed.), *Methods of Molecular Biology, Redox-Mediated Signal Transduction, vol. 476*
© 2008 Humana Press, a part of Springer Science + Business Media, Totowa, NJ
DOI: 10.1007/978-1-59745-129-1_4

for determining intracellular thiol redox status still requires assays of whole-cell extracts to determine ratios of reduced/oxidized redox buffering components. Such an invasive approach has low accuracy and little or no temporal and spatial resolution. The drawbacks of such methods are especially significant, given that individual subcellular compartments often have very different redox environments *(2–4)*. In addition, transient oxidative signaling bursts, such as the localized production of hydrogen peroxide or other highly reactive species, are known to occur but are necessarily restricted to subcellular microdomains *(5, 6)*. Clearly, the study of these complex systems would benefit greatly from use of noninvasive techniques and tools that allow local monitoring of redox status in real time. This chapter focuses on recent developments that allow intracellular redox measurements with high spatial and temporal resolution using oxidation–reduction sensitive variants of the green fluorescent protein (GFP; also *see* Chap. 5 by Mullineaux and Lawson, this volume).

## 2. The Use of GFP and Its Derivatives

### 2.1. The Green Fluorescent Protein

The *Aequorea victoria* GFP is a small protein of 238 amino acids and molecular weight of approx 26 kDa. Its structure is a nearly perfect barrel-shape composed of 11 antiparallel β-strands surrounding a rather distorted coaxial β-helix *(7, 8)*. The "β-can" *(8)* comprising GFP is capped on both ends by short helices and loops that restrict solvent accessibility to the interior of the structure. Three sequential residues on the interior coaxial helix, Ser65, Tyr66, and Gly67, spontaneously form the green light-emitting 4-(*p*-hydroxybenzylidene)imidazolidin-5-one chromophore (**Fig. 4.1**). The nucleophilic attack of the amide nitrogen of Gly67 on the carbonyl carbon of Ser65 forms a five-membered imidazolone ring, followed by dehydration of the carbonyl oxygen of Ser65 and the much-slower step of oxidation of the Tyr66 Cα–Cβ bond to complete the conjugation of the ring systems *(9)*. This process requires no accessory proteins or external cofactors other than molecular oxygen.

GFP is extremely resistant to protease and remains stable under very harsh conditions. In fact, the stability of the protein structure is a requirement for efficient fluorescence, because neither unfolded GFP, nor the naked chromophore, are fluorescent. However, the protein backbone tolerates insertions of entire proteins at several locations, as well as cyclic permutation and/or addition of N- and C-terminal fusion proteins *(10)*. These characteristics make GFP exceptionally suitable as a fusion-tag in both prokaryotic and eukaryotic organisms *(11)*.

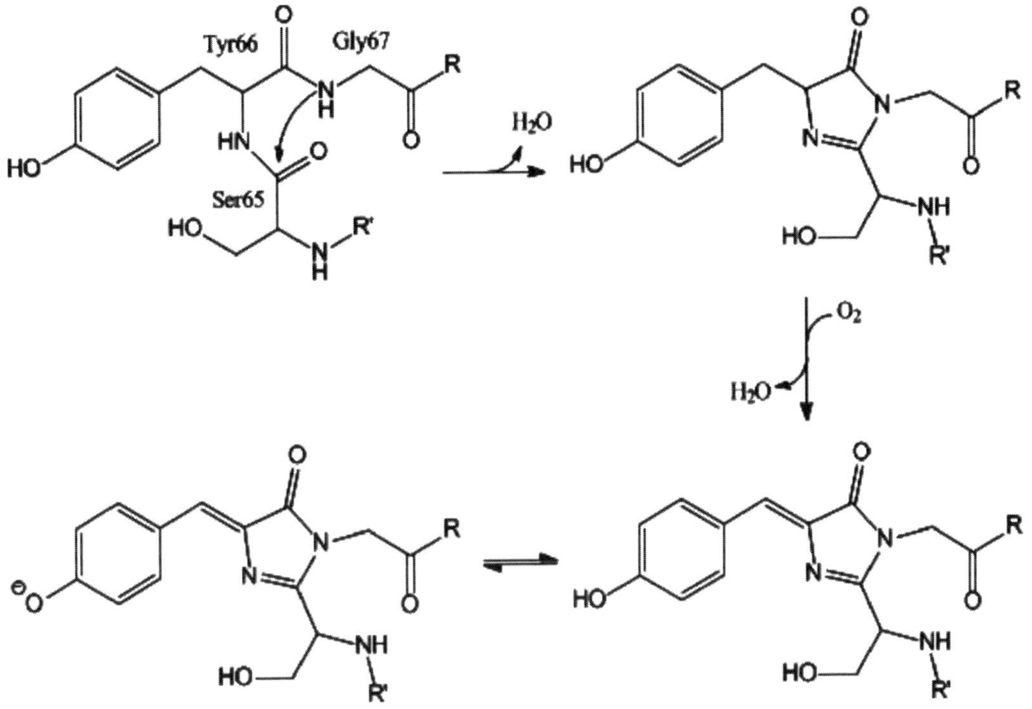

Fig. 4.1. Reaction mechanism for the autocatalytic formation of the GFP chromophore from three amino acid residues in the interior of the β-barrel. The mature chromophore is in equilibrium between the neutral and anionic forms, but excitation of the neutral form leads almost immediately to ESPT and the chromophore, which is then anionic, emits light at 508 nm.

Although several GFP homologs have been isolated from marine organisms, with a rainbow of emission colors ranging from cyan to yellow to red, the original GFP isolated from *Aequorea victoria* remains unique in its complex excitation/emission spectra. Wild-type GFP has two excitation peaks: a major peak at about 400 nm and a minor peak at about 480 nm with approximately one-third the amplitude of the major peak. Excitation at either peak leads to green emission at 508 nm (**Fig. 4.2**). The presence of two separate excitation peaks suggests that within the protein, equilibrium exists between two distinct chromophore states with similar emission wavelengths.

This observation presents a puzzle, which was partially resolved when it was discovered that the two excitation peaks arise from the protonation state of the chromophore. The anionic form of the chromophore is maximally excited at 470–480 nm and emits at about 510 nm. The neutral form of the chromophore, which has a $pK_a$ near neutrality in aqueous solution, is efficiently excited by light with wavelength in the range of 370–400 nm and would normally emit blue light at around 450 nm; however, excited state proton transfer (ESPT) converts the neutral,

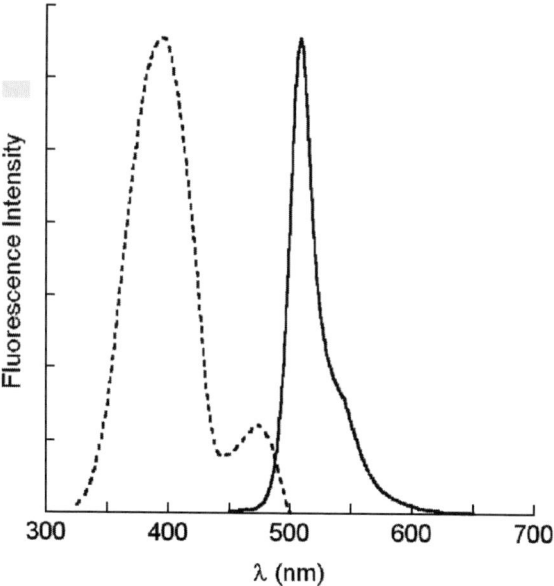

Fig. 4.2. Fluorescence excitation (dashed line, emission recorded at 510 nm) and emission (solid line, excitation at 400 nm) spectra for wild-type GFP. Intensities have been scaled to line up excitation and emission peaks. Excitation at either peak (~400 and 480 nm) leads to the characteristic green emission at 510 nm.

protonated chromophore into an anionic, green emitting species (*see* **Fig.**1 in **ref.** *12*). Upon excitation, the pK$_a$ of the neutral chromophore is dramatically reduced and proton transfer to an acceptor leads to generation of the excited state anion.

Two groups proposed a structural basis for this dual-excitation behavior *(13, 14)*. Upon excitation, the neutral chromophore, which is the dominant ground-state form in wild-type GFP, undergoes rapid ESPT from the Tyr66 phenol via a hydrogen bonding network to internal Glu222. The anionic form of the chromophore, which can also be directly excited at 480 nm, then emits a 510-nm light.

Presumably, the protein structure determines the equilibrium between the neutral and anionic chromophore populations and thus is subject to modification by external influences. Indeed, in the early 1980s it was determined that the equilibrium between protonation states could be perturbed to some extent by changes in pH, salt concentration, or protein concentration (presumably leading to formation of multimers of GFP; *see* **refs.** *15, 16)*. These early results presaged the development of GFP-based indicators of environmental conditions within cells.

The dual excitation behavior of GFP is particularly well suited for the design of *active* biosensors of various cellular phenomena. Through mutagenesis and protein fusions, novel fluorescent protein-based indicators have been developed that respond to a wide variety of compounds and biological events (*see* reviews in **refs**. *17, 18)*. Here, we review the development and application of redox-sensitive GFP indicators.

**2.2. Redox-Sensitive GFP Indicators**

The two widely spaced excitation maxima of wild-type GFP depend on the protonation state of the chromophore, which in turn depends on the structure of the protein. In principle, structural alterations that induce a change in the protonation state form an excellent basis for the creation of *ratiometric* sensors of external conditions. Ratiometric sensors are particularly desirable, because they reduce or eliminate measurement errors due to changes in illumination intensity, cell thickness or indicator concentration.

The technique of ratiometry depends on the presence of two excitation maxima (as in wild-type GFP) or two emission maxima (e.g., red and green), the relative intensities of which are altered in opposite ways by external factors. Assuming two-state behavior, one can factor out the effects of variations in the concentration of the fluorescent indicator and/or the light source intensity by forming a fluorescence intensity ratio. This permits direct quantification of the stimulus *(19)*. On the other hand, an indicator that responds to an environmental stimulus with only an increase or decrease in overall fluorescence may be subject to considerable measurement error because the emission intensity will also depend on variable factors such as illumination intensity, cell thickness, and indicator concentration. In addition, such indicators may be difficult or impossible to calibrate and would give indication only of a relative change in the observed parameter.

Ratiometric redox-sensitive versions of GFP have been developed *(20)* to take advantage of these principles. The probes, termed redox-sensitive green fluorescent proteins (roGFPs), were constructed by placing pairs of cysteine residues on neighboring strands on the surface of the GFP β-barrel in positions favorable for formation of disulfide linkages. Two sites were selected: positions 149/202, and positions 147/204. Six versions of roGFP were initially developed: with cysteine substitutions at the 147/204 site (roGFP1 and roGFP2), at the 149/202 site (roGFP3 and roGFP4), and with cysteine substitutions at both locations (roGFP5 and roGFP6). Half of the roGFPs were developed from wtGFP (roGFP1, roGFP3, and roGFP5) and half from the S65T GFP background (roGFP2, roGFP4, and roGFP6).

The cysteine locations straddle a bulge in the GFP barrel structure around His148, the side chain of which is oriented inside the β-barrel very near the phenolic end of the chromophore (**Fig. 4.3**). Transition between neutral and anionic forms of the chromophore occurs by protonation/deprotonation of this phenolic hydroxyl group, and therefore formation of a strand-bridging disulfide bond at this location was thought to have a

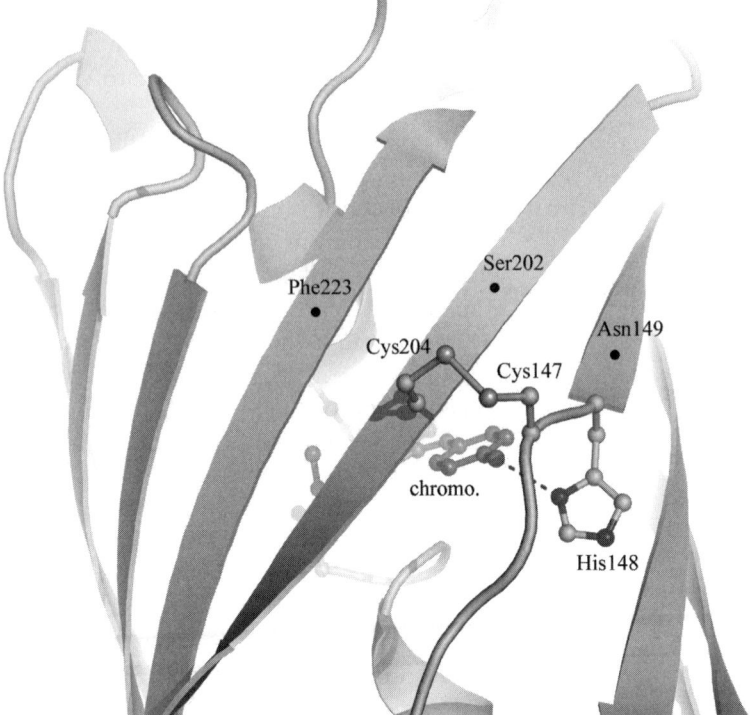

Fig. 4.3. Illustration of roGFP1 (the "R7" variant) in the oxidized form. The chromophore, His148, and the two engineered cysteine residues (Cys147 and Cys204) are shown in ball-and-stick format to illustrate their positions relative to each other. The positions (149, 202, and 223) used for positively charged substitutions in the rate-enhanced variants of roGFP1 are indicated. Figure produced using PyMOL (*43*) and PDB file 2AH8 (*30*).

high probability of affecting fluorescent excitation ratios. A crystal structure of roGFP2 shows that disulfide formation between the pair of engineered surface cysteine residues results in a shift of one β-strand relative to the other *(20)*, which causes subtle internal structural rearrangements, including repositioning of side chains contacting the chromophore (i.e., His148 and Ser205), such that the neutral chromophore is favored over the anionic. Therefore, as a population of roGFP is oxidized, disulfide formation leads to an increase in the excitation peak at 400 nm at the expense of the 480 nm peak *(20)* (*see* **Fig. 4.4**).

RoGFPs were expressed in mammalian cells and were shown to be effective indicators of the ambient cellular redox potential, as perturbed by exogenous oxidants and reductants, as well as by physiological redox changes *(21, 22)*. RoGFPs were expressed in the cytosol as well as in the mitochondrial matrix of HeLa cells. Calibration of the probe was accomplished *in situ* by measuring the 400/480-nm excitation ratio (with emission measured at 508 nm) in the presence of excess exogenous membrane-permeable oxidants ($H_2O_2$) or reductants (dithiothreitol [DTT]).

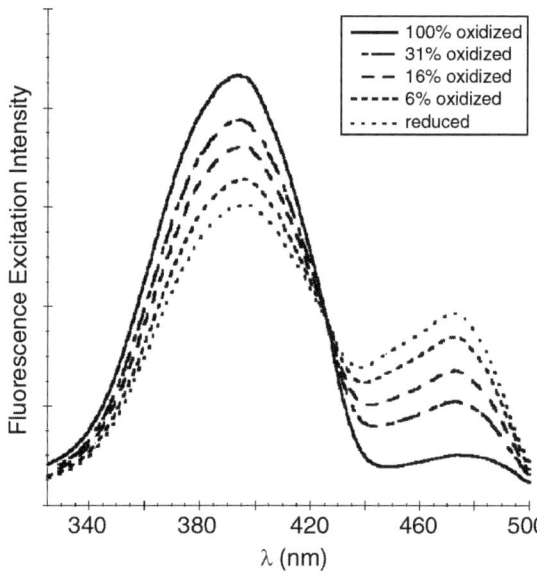

Fig. 4.4. Plot showing a typical redox titration of roGFP1. The *solid line* represents fluorescence excitation from the completely oxidized protein (emission measured at 510 nm). Excitation diminishes at the 400 nm peak while increasing at the 480-nm peak as the protein is reduced. Measurement of 480 to 400 nm excitation peak intensities allows determination of fractional oxidation of the probe, and thus ambient midpoint potential.

These peak ratios correspond to the 100% or 0% oxidized forms of the probe, respectively. In vitro titration of the probe with a range of redox buffers (e.g., oxidized/reduced DTT or GSH/GSSH) and application of the Nernst equation allows alignment of peak ratios, and thus % fractional oxidation of the probe, with redox midpoint potentials (*see* **Fig. 1**, and Materials and Methods in **ref**. *20)*.

The midpoint potentials of oxidation/reduction reactions involving $H^+$ (such as the oxidation/reduction of roGFP) are intrinsically dependent on the pH of the solution. Midpoints calculated using the Nernst equation must therefore by adjusted according to the solution pH. Reduction of a disulfide to the dithiol form requires the input of two $H^+$:

$$roGFP_{ox} + 2H^+ + 2e^- \leftrightarrow roGFP_{red}$$

At the midpoint, $[roGFP_{ox}] = [roGFP_{red}]$ at equilibrium, so the $K_{eq} = [H^+]^{-2}$. The Nernst equation can therefore be used to calculate theoretical midpoints at any pH:

$$E^{o'}_{pH} = E^{o'}_{roGFP} - 0.198T\,(pH-7)\,(mV)$$

Here, $T$ is the absolute temperature in Kelvin and the result is expressed in millivolts. It should be noted that the above reaction assumes both cysteines to be protonated in the reduced form. The pK$_a$s of the engineered cysteine residues in the roGFPs have not been precisely determined, but are most likely > ~9 (MBC and SJR, 2003, unpublished observations), and so at physiological pH this assumption should be valid.

### 2.3. Redox-Sensitive YFP

A similar approach was independently employed by Jakob Winther and co-workers to produce redox-sensitive variants of the GFP-derived yellow fluorescent protein (YFP, [*see* **ref.** *7*]), termed rxYFP *(23)*. The engineered rxYFP cysteines are located at the same position as in roGFP3 and roGFP4 (positions 149/202) and formation of a disulfide bond at this position was shown to affect fluorescence. However, YFP lacks the dual-excitation behavior of GFP and so these probes respond to redox conditions via (non-ratiometric) changes in the amplitude of a single fluorescence excitation peak.

Subsequent studies on rxYFP by Winther et al. *(24, 25)* have revealed important information concerning the cellular interaction partners of GFP-based redox sensors. Thiol redox reactions in cells and subcellular compartments are enzymatically catalyzed, complex, and far from perfectly understood. However, it has been established that the thioredoxin and glutaredoxin systems work independently to equilibrate cellular thiols with different reducing pools, including glutathione and NADP/NADPH *(24)*. Enzyme specificities and reaction kinetics determine which systems interact with specific thiol/disulfides. In yeast, rxYFP apparently equilibrates with the glutathione pool through the actions of glutaredoxins whereas interactions with thioredoxins seem to be much less important in determining the probe's in vivo redox status *(25)*. roGFPs appear to behave similarly (SJR, 2006, unpublished observations).

### 2.4. Re-Engineering the roGFP for Improved Response Rate

One potential application of roGFP is the study of H$_2$O$_2$ bursts in cell signaling events. Recent evidence implicates H$_2$O$_2$ as an important second messenger in cell signaling since it is produced in response to various extracellular stimuli, such as cytokines and peptide growth factors, and its intracellular production or exogenous application affects the function of a variety of proteins, including protein kinases, protein phosphatases, ion channels and transcription factors *(26, 27)*. Cell defense mechanisms, the forefront being the high concentration of reduced glutathione in many cellular compartments *(2, 3, 28)*, quickly eliminate H$_2$O$_2$ and other potentially damaging reactive oxygen species and therefore such oxidative bursts are believed to be very transient and highly localized. Understanding the complex relationships involved in this aspect of cell signaling will require specific probes with high temporal and spatial resolution. Dooley et al. *(22)* investigated the possibility of employing roGFP to detect these oxidative bursts, but were unable to measure any significant changes in roGFP excitation peak ratios in response to intracellular $H_2O_2$ production *(22)*. It is likely that this inadequacy is due to the relatively slow response time of the roGFP probes, which are on the order of tens of minutes.

Formation of a disulfide bond requires deprotonation of the cysteine thiol to the thiolate form. Once this rate-limiting step has occurred, disulfide formation rapidly follows. Stabilization of the cysteine thiolate by nearby positive charges or dipoles is thought to lower the activation energy of the process and increase disulfide formation rates *(29)*. One approach to improving response time of such a system is to substitute positively charged residues near the reactive thiols, thus lowering their $pK_a$s and increasing disulfide formation rates. Using this strategy, several variants of roGFP1 (designated roGFP1-R1 through roGFP1-R14) have been constructed with up to three positively charged substitutions (lysine or arginine) at locations near the reactive cysteines, positions 149, 202, and 223 *(30)* (*see* **Fig. 4.3**). Substantial rate increases were observed. Mutants with a single basic substitution exhibited an approximate doubling of the pseudo first-order rate constant, as measured by monitoring excitation peak ratios over time after the addition of excess DTT or $H_2O_2$ to buffered solutions of the roGFP1 variants. Each additional basic substitution further increased the rate by approximately twofold, with a maximum rate increase of approximately sevenfold in the variant with three basic substitutions.

Application of nonlinear Poisson-Boltzmann theory (as implemented in the program DelPhi; *see* **refs. 31, 32)** verified $pK_a$ depression of the reactive cysteine thiols. Other groups have undertaken similar approaches to increase cysteine reactivity, with comparable results *(22, 33)*. Although the observed rate increases from this approach are significant, the solvent-exposed location of these substitutions on the outer surface of a β-barrel imposes natural limits on its effectiveness since these engineered charge-charge interactions are medium-range at best (*see* **Fig.** 3 of **ref.** *30)*. Nevertheless, the roGFP1 variant with the most significant rate enhancement (termed "roGFP1-R12") is recommended for general use because of its faster response time as well as its more oxidizing disulfide midpoint potential.

### 2.5. Midpoint Potentials of Redox Probes

Another potential limitation on the use of roGFP probes concerns the rather negative midpoint potentials. RoGFPs 1–6 exhibit midpoint potentials that range from –272 mV (roGFP2) to –299 mV (roGFP3) *(20)*. The utility of the probes in determining midpoint potential drops off rapidly the farther the ambient midpoint is from the probe's own midpoint potential. In practice, roGFPs are most useful in measuring midpoints within ~35 mV of their own midpoint (*see* **Fig.** 1 in **ref.** *20)*. RoGFP1, for example, has a measurement range of ~ –325 to –255 mV, (i.e., when the probe is between ~10% and 90% oxidized). This range makes roGFP1 (and its siblings) ideal for determination of thiol redox status in highly reducing compartments, such as the cytosol or mitochondria. However, in more oxidizing (i.e., more positive midpoint potential) compartments such as the endoplasmic reticulum (ER), the probes are expected to be completely oxidized and thus capable of indicating only this fact.

Re-engineered roGFPs such as roGFP-R12 are somewhat more oxidizing than the first-generation probes. The introduced basic residues act to lower the cysteine thiol p$K_a$s, thus stabilizing the thiolate anion, and this, in general, destabilizes the disulfide thermodynamically *(34, 35)*. RoGFP1-R variants with more basic substitutions had slightly more oxidizing midpoints, with roGFP1-R12 and roGFP1-14 (each with three basic mutations near the reactive cysteines) having the most oxidizing midpoint potentials, –265 and –263 mV, respectively *(30)*. However, the cysteine p$K_a$s remain high, greater than 9, and it would be desirable to lower these values substantially (*see* below).

## 3. Recent Applications of roGFPs

### 3.1. Plant Cell Expression

Jiang and co-workers have recently transformed *Arabidopsis* with roGFP1, targeted to both the mitochondria and cytosol *(36)*. Intracellularly expressed roGFP1 responded to the addition of exogenous oxidizing and reducing agents in a like manner to that reported for mammalian cells and in vitro assays *(20, 22)*; therefore, the same calibration curves were used to convert percent oxidation of the probe to redox potentials. Parallel monitoring of the redox potential with an external redox electrode confirmed the response of roGFP1 fluorescence excitation ratios to changes in redox status. The resting redox potential of the cytoplasm, measured at three different root zones, was found to be –318 ± 13 mV, a value significantly more oxidizing than that determined from roGFP1 expressed in the mitochondria (–362 ± 10 mV). These values are nearly identical to those reported for mammalian cells (–360 mV for mitochondria in HeLa cells *(20)*, and –315 mV reported for HeLa cell cytosol *(22)*). Jiang et al. also reported a more robust redox buffering capacity for the mitochondria over the cytosol, measured by comparing changes in percent oxidation of the probe in both compartments after the addition of $H_2O_2$. A comparison of the resting redox state of three root zones (root cap, meristem, and zone of elongation) revealed a somewhat-more oxidizing state in the zone of elongation compared with the meristem as well as a smaller redox buffering capacity, consistent with reports that ROS may be used as second messengers in plant cells to signal certain kinds of growth *(37)*.

**3.2. Monitoring Oxidative Stress in Ischemic Neuronal Cells**

Ischemia in neurons leads to cell-damaging oxidative stress and sensitivization involving depletion of glutathione levels in a process that is not yet well understood. Glutathione acts as an antioxidant buffering pool against ROS. Glutaredoxins, which derive their reducing equivalents from glutathione are, along with thioredoxins, the principle enzymes that function to maintain cytosolic thiols in their reduced state. Consequently, the depletion of glutathione pools would limit the cell's ability to respond adequately to oxidative stress. In a study of the consequences of ischemia-induced oxidative stress, Vesce and co-workers expressed roGFP2 in neuronal cells to confirm that depletion of glutathione would result in a more oxidizing thiol redox potential *(38)*. Using monochlorobimane (mBCl), which depletes glutathione by forming a fluorescent conjugate, the researchers were able to mimic this consequence of ischemia. RoGFP2, expressed in these neurons, became significantly oxidized upon the application of mBCl. Unfortunately, fluorescence from the mBCl-glutathione conjugate interferes with measurement of roGFP emission from 400-nm excitation, thus, probe response had to be measured by comparing excitation intensities at 480 nm instead of excitation ratios.

**3.3. Measurement of Endosomal Redox Potential**

Antibody–drug conjugate therapy is a process whereby cytotoxic compounds are specifically targeted to cancerous cells in order to maximize the effectiveness of chemotherapeutic agents while minimizing harm to the noncancerous cells of the patient. The drug is directed to tumor cells via a covalent linkage to an antibody targeting a tumor-specific antigen. Several types of covalent linkages have been used between drug and antibody, including disulfide bonds. Disulfide linkages have been shown to be effective in delivering toxic molecules to the interior of a tumor cell *(39, 40)*, but it had been incorrectly assumed that the effect was due to a reducing environment during internalization. Scales and co-workers *(4)*, in the process of investigating the intracellular cleavage of these disulfide-liked antibody–drug conjugates, found surprising evidence that endosomal compartments are not reducing but quite oxidizing, with midpoint potentials similar to those found in the ER *(4)*.

Austin et al. *(41)* developed a novel intracellular cleavage assay using the drug trastuzumab attached to the fluorophore Rhodamine red (RR) via a disulfide-based linker. Trastuzumab had previously been shown to be effective against breast cancer tumors because it is an antibody specific for HER2, a member of the epidermal growth factor receptor family of tyrosine kinases that is overexpressed in many tumor cells. HER2 is cell-surface localized but, interestingly, undergoes endocytosis at the rate of 1–2% per min, followed by recycling back to the cell surface *(41)*. Upon endosomal cleavage of the disulfide-based trastuzumab–RR linker, there should be detectable fluorescence dequenching from the now free RR. However, because very little dequenching was observed, the researchers questioned the assumption of a reductive milieu in the endosomal compartments.

Before the development of roGFPs, no suitable redox probe was available with the pH- and proteolysis-resistance necessary to investigate conditions within endosomes. RoGFP1 was therefore targeted to various endosomal compartments in order to examine their redox status. In control experiments, roGFP1 was also targeted to the mitochondria and ER of the same cells. Measurements of the 380/490 nm excitation peak ratios of these constructs confirmed the suspicion that endosomal compartments are oxidizing, rather than reducing. Mitochondrial roGFP was 8.3% (±5.4%) oxidized, in agreement with that observed in mammalian cells *(20, 22)*, whereas ER-targeted roGFP was 95.8% (±9.7%) oxidized *(4)*. Endosomally targeted roGFPs, however, were 94–97% oxidized. Such high levels of roGFP oxidation prohibit precise determination of midpoint potentials, but Austin et al. *(41)* estimated that these compartments are at least as oxidizing as –240 mV, and far more oxidizing than observed for mitochondria.

## 4. Future Directions

Two major limitations of redox-sensitive GFPs remain to be resolved. First, the response time, although notably improved in the re-engineered roGFPs, likely remains too slow to permit accurate indication of transient oxidative bursts involved in some types of signaling events. Second, the highly reducing midpoint potentials of roGFP probes currently limit quantitative applications to reducing compartments such as the cytosol and mitochondria. Strategies to develop roGFP variants with more oxidative midpoints may include the insertion or deletion of residues along the cysteine containing β-strands so as to increase geometric strain in the disulfide. It is commonly assumed (with much evidence in support) that disulfide midpoint potential is strongly influenced by factors such as electrostatic modification of cysteine $pK_a$ and geometric strain in the disulfide linkage; however, the relationships are not well understood *(30)*. Furthermore, there may be inherent limitations in the fold of the GFP molecule that restrict one's freedom to re-engineer the protein molecule.

Cysteine substitutions at other positions on the surface of the GFP β-barrel may allow the introduction of fluorescence-modifying disulfides with different stabilities. Alternatively, protein fusions capable of intramolecular disulfide formation, or other more dramatic modifications might result in a reactive disulfide contained within a more protected pocket, similar to the reactive disulfides found in thioredoxins or peroxiredoxins. For example, members of the thioredoxin superfamily have catalytic cysteine residues with $pK_a$s < 4, and midpoint potentials as high as –122 mV *(35, 42)*. Such a protected, low dielectric location would allow greater contributions from nearby charges that could alter cysteine thiol $pK_a$s to more closely resemble these enzymes' active thiolates, and thus dramatically increase response time and midpoint potential.

## 5. Conclusions

RoGFPs were originally designed to permit one to conduct non-invasive quantification and/or comparison of thiol/disulfide equilibrium in specific intracellular compartments. As the aforementioned preliminary examples illustrate, those goals have not only been met, but some of the results have already challenged conventional thinking in surprising ways. Furthermore, roGFP probes respond to a variety of oxidative events and are in principle capable of reporting the transient generation of reactive oxygen species. It was surprising to us, but also very encouraging, that the development of a probe to quantify redox conditions within a cell could lead to new techniques for the study of cell signal transduction.

## References

1. Jacob, C., Knight, I., and Winyard, P. G. (2006) Aspects of the biological redox chemistry of cysteine: from simple redox responses to sophisticated signalling pathways. *Biol Chem* **387**, 1385–1397.

2. Hwang, C., Sinskey, A. J., and Lodish, H. F. (1992) Oxidized redox state of glutathione in the endoplasmic reticulum. *Science* **257**, 1496–1502.

3. Schafer, F. Q. and Buettner, G. R. (2001) Redox environment of the cell as viewed through the redox state of the glutathione disulfide/glutathione couple. *Free Radic Biol Med* **30**, 1191–1212.

4. Austin, C. D., Wen, X., Gazzard, L., Nelson, C., Scheller, R. H., and Scales, S. J. (2005) Oxidizing potential of endosomes and lysosomes limits intracellular cleavage of disulfide-based antibody–drug conjugates. *Proc Natl Acad Sci USA* **102**, 17987–17992.

5. Rhee, S. G., Chang, T. S., Bae, Y. S., Lee, S. R., and Kang, S. W. (2003) Cellular regulation by hydrogen peroxide. *J Am Soc Nephrol* **14**, S211–S215.

6. Rhee, S. G., Bae, Y. S., Lee, S. R., and Kwon, J. (2000) Hydrogen peroxide: a key messenger that modulates protein phosphorylation through cysteine oxidation. *Sci STKE* **2000**, PE1.

7. Ormo, M., Cubitt, A. B., Kallio, K., Gross, L. A., Tsien, R. Y., and Remington, S. J. (1996) Crystal structure of the Aequorea victoria green fluorescent protein. *Science* **273**, 1392–1395.

8. Yang, F., Moss, L. G., and Phillips, G. N. J. (1996) The molecular structure of green fluorescent protein. *Nat Biotechnol* **14**, 1246–1251.

9. Heim, R., Prasher, D. C., and Tsien, R. Y. (1994) Wavelength mutations and post-translational autoxidation of green fluorescent protein. *Proc Natl Acad Sci USA* **91**, 12501–12504.

10. Baird, G. S., Zacharias, D. A., and Tsien, R. Y. (1999) Circular permutation and receptor insertion within green fluorescent proteins. *Proc Natl Acad Sci USA* **96**, 11241–11246.

11. Tsien, R. Y. (1998) The green fluorescent protein. *Annu Rev Biochem* **67**, 509–544.

12. Chattoraj, M., King, B. A., Bublitz, G. U., and Boxer, S. G. (1996) Ultra-fast excited state dynamics in green fluorescent protein: multiple states and proton transfer. *Proc Natl Acad Sci USA* **93**, 8362–8367.

13. Brejc, K., Sixma, T. K., Kitts, P. A., Kain, S. R., Tsien, R. Y., Ormo, M., and Remington, S. J. (1997) Structural basis for dual excitation and photoisomerization of the Aequorea victoria green fluorescent protein. *Proc Natl Acad Sci USA* **94**, 2306–2311.

14. Palm, G. J., Zdanov, A., Gaitanaris, G. A., Stauber, R., Pavlakis, G. N., and Wlodawer, A. (1997) The structural basis for spectral variations in green fluorescent protein. *Nat Struct Biol* **4**, 361–365.

15. Ward, W. W. and Bokman, S. H. (1982) Reversible denaturation of Aequorea green-fluorescent protein: physical separation and characterization of the renatured protein. *Biochemistry* **21**, 4535–4540.

16. Ward, W. W., Prentice, H. J., Roth, A. F., Cody, C. W., and Reeves, S. C. (1982) Spectral perturbations of the Aequorea green-fluorescent protein. *Photochem Photobiol* **35**, 803–808.

17. Giepmans, B. N., Adams, S. R., Ellisman, M. H., and Tsien, R. Y. (2006) The fluorescent protein toolbox for assessing protein location and function. *Science* **312**, 217–223.

18. Zhang, J., Campbell, R. E., Ying, A. Y., and Tsien, R. Y. (2002) Creating new fluorescent probes for cell biology. *Nat Rev Cell Biol* **3**, 906–918.

19. Grynkiewicz, G., Poenie, M., and Tsien, R.Y. (1985) A new generation of Ca2+ indicators with greatly improved fluorescence properties. *J Biol Chem* **260**, 3440–3450.

20. Hanson, G. T., Aggeler, R., Oglesbee, D., Cannon, M., Capaldi, R. A., Tsien, R. Y., and Remington, S. J. (2004) Investigating mitochondrial redox potential with redox-sensitive green fluorescent protein indicators. *J Biol Chem* **279**, 13044–13053.

21. Rossignol, R., Gilkerson, R., Aggeler, R., Yamagata, K., Remington, S. J., and Capaldi, R. A. (2004) Energy substrate modulates mitochondrial structure and oxidative capacity in cancer cells. *Cancer Res* **64**, 985–993.

22. Dooley, C. T., Dore, T. M., Hanson, G. T., Jackson, W. C., Remington, S. J., and Tsien, R. Y. (2004) Imaging dynamic redox changes in mammalian cells with green fluorescent protein indicators. *J Biol Chem* **279**, 22284–22293.

23. Ostergaard, H., Henriksen, A., Hansen, F. G., and Winther, J. R. (2001) Shedding light on disulfide bond formation: engineering a redox switch in green fluorescent protein. *EMBO J* **20**, 5853–5862.

24. Bjornberg, O., Ostergaard, H., and Winther, J. R. (2006) Measuring intracellular redox conditions using GFP-based sensors. *Antioxid Redox Signal* **8**, 354–361.

25. Ostergaard, H., Tachibana, C., Winther, J. R. (2004) Monitoring disulfide bond formation in the eukaryotic cytosol. *J Cell Biol* **166**, 337–345.

26. Rhee, S. G., Kang, S. W., Jeong, W., Chang, T. S., Yang, K. S., and Woo, H. A. (2005) Intracellular messenger function of hydrogen peroxide and its regulation by peroxiredoxins. *Curr Opin Cell Biol* **17**, 183–189.

27. Finkel, T. (1998) Oxygen radicals and signaling. *Curr Opin Cell Biol* **10**, 248–253.

28. Tu, B. P., Ho-Schleyer, S., Travers, K. J., and Weissman, J. S. (2000) Biochemical basis of oxidative protein folding in the endoplasmic reticulum. *Science* **290**, 1571–1574.

29. Kim, J. R., Yoon, H. W., Kwon, K. S., Lee, S. R., and Rhee, S. G. (2000) Identification of proteins containing cysteine residues that are sensitive to oxidation by hydrogen peroxide at neutral pH. *Ana. Biochem* **283**, 214–221.

30. Cannon, M. B., Remington, S. J. (2006) Re-engineering redox-sensitive green fluorescent protein for improved response rate. *Protein Sci* **15**, 45–57.

31. Rocchia, W., Alexov, E., and Honig, B. (2001) Extending the applicability of the nonlinear Poisson–Boltzmann equation: multiple

dielectric constants and multivalent ions. *J Phys Chem B* **105**, 6507–6514.

32. Rocchia, W., Sridharan, S., Nicholls, A., Alexov, E., Chiabrera, A., and Honig, B. (2002) Rapid grid-based construction of the molecular surface for both molecules and geometric objects: applications to the finite difference Poisson-Boltzmann method. *J Comp Chem* **23**, 128–137.

33. Hansen, R. E., Ostergaard, H., and Winther, J. R. (2005) Increasing the reactivity of an artificial dithiol-disulfide pair through modification of the electrostatic milieu. *Biochemistry* **44**, 5899–5906.

34. Creighton, T. E. (1975) Interactions between cysteine residues as probes of protein conformation: the disulfide bond between Cys-14 and Cys-38 of the pancreatic trypsin inhibitor. *J Mol Biol* **96**, 767–776.

35. Nelson, J. W. and Creighton, T. E. (1994) Reactivity and ionization of the active site cysteine residues of DsbA, a protein required for disulfide bond formation *in vivo*. *Biochemistry* **33**, 5974–5983.

36. Jiang, K., Schwarzer, C., Lally, E., Zhang, S., Ruzin, S., Machen, T., Remington, S. J., and Feldman, L. (2006) Expression and characterization of a redox-sensing green fluorescent protein (reduction-oxidation-sensitive green fluorescent protein) in Arabidopsis. *Plant Physiol* **141**, 397–403.

37. Joo, J. H., Bae, Y. S., and Lee, J. S. (2001) Role of auxin-induced reactive oxygen species in root gravitropism. *Plant Physiol* **126**, 1055–1060.

38. Vesce, S., Jekabsons, M. B., Johnson-Caldwell, L. I., and Nicholls, D. G. (2005) Acute glutathione depletion restricts mitochondrial ATP export in cerebellar granule neurons. *J Biol Chem* **280**, 38720–38728.

39. Calvete, J. A., Newell, D. R., Wright, A. F., and Rose, M. S. (1994) *In vitro* and *in vivo* antitumor activity of ZENECA ZD0490, a recombinant ricin A-chain immunotoxin for the treatment of colorectal cancer. *Cancer Res* **54**, 4684–4690.

40. Liu, C., Tadayoni, B. M., Bourret, L. A., Mattocks, K. M., Derr, S. M., Widdison, W. C., Kedersha, N. L., Ariniello, P. D., Goldmacher, V. S., Lambert, J. M., Blattler, W. A., and Chari, R. V. (1996) Eradication of large colon tumor xenografts by targeted delivery of maytansinoids. *Proc Natl Acad Sci USA* **93**, 8618–8623.

41. Austin, C. D., De Maziere, A. M., Pisacane, P. I., van Dijk, S. M., Eigenbrot, C., Sliwkowski, M. X., Klumperman, J., and Scheller, R. H. (2004) Endocytosis and sorting of ErbB2 and the site of action of cancer therapeutics trastuzumab and geldanamycin. *Mol Biol Cell* **15**, 5268–5282.

42. Aslund, F. and Beckwith, J. (1999) The thioredoxin superfamily: redundancy, specificity and gray-area genomics. *J Bacteriol* **181**, 1375–1379.

43. DeLano, W. L. (2002) The PyMOL Molecular Graphics System on World Wide Web: http://www.pymol.org.

# Chapter 5

# Measuring Redox Changes In Vivo in Leaves: Prospects and Technical Challenges

## Philip M. Mullineaux and Tracy Lawson

## Abstract

In leaves, the functioning of many key proteins under conditions promoting oxidative stress depends to a large extent on the redox potential of the glutathione couple. Routine measurements of the glutathione pool in leaves are destructive and labor-intensive processes that tend to underestimate the redox state. Therefore, a challenge for plant scientists is to develop a tool capable of measuring the redox state of the glutathione couple spatially (at different levels of resolution) and temporally in tissues and subcellular compartments in vivo. This chapter highlights the possibilities of using redox-sensitive green fluorescence proteins (roGFPs) as real-time redox reporters for use in intact plants and focuses on practical assessments of using such bioindicators in different leaf cell types subjected to environmental change. The advantages and shortcomings of different GFP variants are discussed along with the choice of system for leaves and possible approaches to overcoming some of the problems. We consider roGFP1-12 as an ideal candidate for developing a redox reporter system in whole plants because it has several advantages over the other variants, with dual excitation peaks allowing a ratiometric approach, insensitivity to pH and halide ions, increased response times for real-time measurements, and appropriate emission wavelengths for use in leaves. We conclude that when using roGFP1-12 with specific cell promotors, it would be possible to target distinct cell compartments and tissues and monitor changes in glutathione redox state to determine the effects of reactive oxygen species on specific cellular components.

**Key words:** Intact leaves, glutathione pool, redox state, reporter, roGFP.

## 1. Introduction

In plant cells, as in all other living cells, most of metabolism takes place in a reducing environment in subcellular compartments, such as the cytosol and the chloroplast stroma, because the many enzymes that have critical thiol groups as part of their catalytic

John T. Hancock (ed.), *Methods in Molecular Biology, Redox-Mediated Signal Transduction, vol. 476*
© 2008 Humana Press, a part of Springer Science+Business Media, Totowa, NJ
DOI: 10.1007/978-1-59745-129-1_5

sites require protection from oxidation *(1)*. However, some sub-cellular compartments, in particular the endoplasmic reticulum *(2)*, may be more oxidizing to promote folding of proteins via the formation of disulfide bridges *(3)*. When oxidative stress occurs, the reducing state of the cell can be overwhelmed, resulting in cell death *(4)*. Despite the potential for oxidative damage to plant cells, transitory changes in reactive oxygen species (ROS), such as hydrogen peroxide ($H_2O_2$), superoxide anion ($O_2^-$), and singlet oxygen ($^1O_2$) occur, which have been suggested to initi-ate signaling for defence against pathogens *(5)*, acclimation to a range of abiotic stresses *(1)*, and developmental processes, such as secondary root hair formation and xylem differentiation *(6)*. Changes in ROS may be associated with a transitory decrease in the redox status of some subcellular compartments, such as the apoplast during induction of defence against pathogens and the chloroplast stroma during exposure of plants to high light *(7, 8)*. However, it has been argued that such changes in ROS levels associated with signaling responses are limited to discrete parts of the cell contained by a reducing environment that may be refrac-tory to changes in its redox status *(4, 9)*.

In subcellular compartments, such as the cytosol, the pres-ence of millimolar amounts of reduced glutathione (GSH) have been suggested to act as a cellular redox buffer keeping many proteins cysteine moieties in their reduced state (e.g., Noctor et al. 2002; Mullineaux and Rausch 2005) *(10, 11)*. This view is undergoing modification as a result of recent research of authors who suggest that a better way of viewing the role of GSH and its oxidized form, glutathione disulphide (GSSG), both directly and as an enzyme co-factor, is to ensure that redox active protein thiols are kept in specific oxidation states and prevented from entering terminal, irreversible oxidation states *(3)*. Given the millimolar concentration of GSH in plant cells *(10)*, it is of con-siderable importance to many biologists to have reliable methods for measuring the in vivo redox potential of the glutathione redox couple since it is likely to impact on the functioning of many key proteins under conditions that promote oxidative stress.

## 2. Limitations and Possibilities

The challenge for plant scientists is to measure the redox state of the GSH couple spatially (at different levels of resolution) and temporally in tissues of plants of different ages as they respond to the multiple cues in their environment. Detailed methods for measuring redox changes in real time in cells are only just becoming established for single-cell systems (also *see* Chap. 4), and measurements of redox changes in cells of multicellular organisms are just beginning *(12)*. This chapter is unusual in the context of this book, because we have not produced a set of methodologies but instead have focus on a practical assessment of the available green fluorescent protein (GFP)-based redox sensors for their potential to inform us about real-time GSH redox changes in different cell types in the mature intact leaf subject to environmental change. To date, and to our knowledge, this has not been reported. Therefore, we have set out to briefly summarize the key features of redox-sensitive GFPs and to consider the development of a redox-sensitive GFP reporter system for the specific technically demanding application to a mature leaf of a plant grown in soil.

*2.1. Determination of GSH Levels and Redox Status in the Leaves of Plants*

Most of the data available from the measurements of GSH and GSSG in leaves require that the leaves themselves be destroyed, are labor intensive, and have a tendency to underestimate GSH redox state, principally because of problems of determining relatively low levels of GSSG. Values in the order of 200–300 nmol g/fwt are typical for unstressed plants *(10)*.

Some success has been achieved in the measurement of cytosolic GSH concentrations $[GSH_{cyt}]$ in situ in leaf tissues with the use of the GSH-specific compound monochlorobimane (MCB), which forms a glutathione (GS)–MCB fluorescent adduct in a glutathione-*S*-transferase (GST)-catalysed reaction *(13)*. MCB is infiltrated into leaves before detection, and imaging of fluorescence is conducted with the use of laser confocal microscopy. As well as extensive work conducted on roots and suspension culture cells *(10)*, estimates of 200–300 $\mu M$ for a range of photosynthetic and nonphotosynthetic (epidermal cells) of poplar leaves was reported *(14)* and in trichome basement and epidermal cells of *Arabidopsis* leaves 238, 80, and 144 $\mu M$ were determined, respectively *(15)*.

However, this method suffers several serious drawbacks, including the fact that no measurement of the GSH-to-GSSG ratio to calculate redox state is possible, the delivery of an excess of MCB may deplete GSH, leading to adverse physiological responses, and the requirement to inhibit export of GS–MCG adducts into the vacuole using azide *(13)*. Finally, the reliance upon GST activity in the compartment to be measured means that if there is no GST present (most notably in the chloroplast; see ref. *14*), then no determination can be made.

# 3. Redox GFPs

**Section 2** highlights the importance of a reporter protein-based real time redox assay. The advantages of using GFPs and its color variants (e.g., yellow fluorescent proteins) for a variety of applications and how the chromophore is formed have been reviewed previously (*see* ref. *16*, and Chap. 4).

In the context of the use of GFP for redox-sensing in vivo, the opportunity to conduct noninvasive experiments with a protein which is regarded as neutral in its impact on biological systems, contains no disulfide bonds, can be expressed in a tissue-specific manner and be targeted to specific subcellular compartments makes this an attractive starting point (*17*). However, in plant cells transformed with GFP-based genes, high-level expression (i.e., the strongest fluorescers) is sometimes not recovered, suggesting that deleterious effects on plant cells can occur (*18*). Thus, at lower levels of expression of GFP, more subtle perturbations may operate, which the researcher ought to keep in mind. Nevertheless, the potential advantages of a noninvasive system have spurred the independent development of two types of redox GFPs; redox-sensitive yellow fluorescent proteins (rxYFPs; ref. *17*) and redox GFPs (roGFPs; ref. *2*).

### 3.1. A Comparative Description of the rxYFP and roGFP Redox Reporter Systems

Both systems, although developed independently, have several features in common, but also some differences, which will influence the choice of which system to deploy for leaves of higher plants. Both types of GFP have been made redox-sensitive by the introduction of solvent-exposed cysteine residues by site-directed mutagenesis into adjacent strands of the β-barrel that surrounds the GFP chromophore. Upon oxidation of the redox reporters, the principle behind the creation of a disulfide bridge in the β-barrel is to bring about a conformational change that perturbs the structure of the chromophore and favors its protonated (neutral) form over its anionic form, bringing about a change in the fluorescence properties of the GFPs. This feature is important for considerations on the choice of rxYFP or roGFP for leaves.

In both initial reports describing rxYFP and roGFP, the same β strands of the protein were targeted. For the rxYFP variant protein harboring the mutations N149C and S202C, the difference between maximal fluorescence of the reduced form and diminished fluorescence of the oxidized form was the greatest (*17*) whereas in the roGFP variants produced (roGFP1 and roGFP2), the mutations were S147C and Q204C (*2*).

Despite the introduction of similarly positioned disulfide bridge, these proteins do have different midpoint redox potentials and therefore may well be in equilibrium with different cellular redox couples. RxYFP has a midpoint redox potential ($E_0$) of −261 mV, determined by titration against a range of GSH to GSSG ratios (*17*) and was suggested to accurately report the redox status of the glutathione pool in vivo. This was subsequently confirmed in yeast with the use of mutants with altered GSH levels and GSH-to-GSSG ratios (*19*). In wild-type *Escherichia coli* and yeast cells, it was estimated that steady-state oxidized protein levels were present at approx 50% and 90%, respectively, of the total amount of rxYFP present *(17, 19)*. One problem with the rxYFP system is that the best fluorescence difference that could be obtained between oxidised and reduced form was approx. 2.2-fold, which possibly is a limitation for measuring subtle changes in cellular redox potential *(17)*.

In contrast, the roGFP1 and roGFP2 variant proteins have consensus midpoint redox potentials of –291 and –287 mV, respectively, using redox titrations with dithiothreitol, lipoic acid, and bis(2-mercaptoethyl)sulfone redox couples *(2, 20)*. The highly reducing redox potential of roGFPs makes it unclear as to the cellular redox couple(s) the protein would be in equilibrium with. In addition, the rates of changes in redox potential are in the order of tens of minutes *(21)* making them less than ideal to respond to transitory changes in redox states associated with the production of short bursts of ROS.

### 3.2. Improved Version of Redox-GFPs

The original roGFPs and rx YFPs have been modified further to address some of the deficiencies in the original system. In both cases, changing the electrostatic environment by introducing positively charged amino acids around the disulfide bridge improved the properties of redox reporters. In the case of roGFP1, the changes N149K, F223R, and S202K (roGFP1-R12 variant) were particularly effective at lowering the midpoint redox potential of the reporter to –265 mV and increasing the rates of reduction and oxidation of fivefold and sixfold, respectively *(21)*. In the case of rxYFP, the combined changes of Y200R, Q204R, and A227R resulted in a 13-fold difference in fluorescence between its maximal oxidised and reduced forms *(22)*.

### 3.3. Ratiometric Versus Absolute Determinations of Fluorescence

One of the crucial differences between the two types of redox reporters is the original choice of starting fluorescent protein and the consequences this has for the use of these proteins as redox reporters in organs such as the leaf. Wild-type GFP from *Aequoria victoria* exhibits two excitation maxima at wavelengths of 400 and 475–490 nm. These maxima correspond to absorption of the anionic and neutral forms of the chromophore *(23)*. In wild-type GFP, both of these excitation maxima elicit fluorescence (maximum at 508 nm). However, in the production of "color" forms of GFP, including YFP, the dual excitation maxima have been replaced by a single "red-shifted" excitation and emission maximum *(23)*. In roGFP1, derived from wild-type GFP *(2)*, the dual excitation maxima are retained and, as the protein becomes more oxidized, there is a decrease in the fluorescence from the 400-nm excitation and an increase in the fluorescence from the 470-nm excitation *(2, 20)*. This means that a ratiometric approach to measuring changes in fluorescence can be adopted in which the fluorescence values can be expressed as a ratio of the excitation at 400/470 nm, thus resulting in an increase in the ratio as roGFP1 becomes more oxidized, the maximum difference in ratios between fully oxidized and reduced forms of roGFP1 being sixfold in vitro *(2, 20, 21)*. Ratiometric determinations of fluorescence eliminate or reduce possible artefactual influences on the data caused by photobleaching or inadvertent modification of the protein, protein concentration, illumination stability, excitation path length and non-uniform distribution of the protein in groups of cells *(2, 20)*.

Determination of redox potential using rxYFP relies upon direct measurement of fluorescence quenching as the protein becomes more oxidised and vice versa when being reduced *(17)*. The ratiometric approach to collection of fluorescence data is not an option for this system, but nevertheless with appropriate estimations of protein concentrations, this system has proved very effective in the experimental systems where it has been reported so far *(17, 19)*.

A further problem that might also dictate choice of redox probe is that YFP fluorescence and its variants can be quenched by anions, in particular halide ions, and is more susceptible to pH change. This need not be a problem in some experimental systems but it should be noted that roGFP1 is relatively insensitive to these problems *(2, 20)*.

### 3.4. Monitoring Cellular Redox Status In Vivo Using rxYFP and roGFP1

Both version of redox GFPs have been used successfully in living cells. RxYFP has been used in *E. coli* and yeast *(17, 19)*, whereas roGFPs have been targeted to the cytosol and mitochondria in mammalian cells and in transgenic *Arabidopsis* and measured in roots *(2,12, 20)*.

A clear demonstration of the potential of redox reporters to provide important information about the glutathione redox couple in cells was first shown for the cytosol of *E. coli* harboring rxYFP and the impact of the loss of the thioredoxin system for maintaining reduced proteins. Mutants deficient in thioredoxin (*trxB*) showed 30% more oxidized rxYFP than wild-type cultures. Thioredoxin does not react with rxYFP and the cytoplasmic GSH/GSSG redox potential and concentration was close to the value estimated in vitro for when the rxYFP and glutathione redox couples were in equilibrium. Therefore, these data suggested that loss of the thioredoxin system impacts on the redox state of the glutathione pool, which partly compensates for the loss of the thioredoxin system.

It should be noted that corrections for the amount of reduced and oxidized rxYFP have to be performed in both the bacteria and yeast experiments *(17, 19)* to ensure that fluorescence yield was not a consequence of variation in the amount of protein. This was achieved by extracting cultures directly into strong acid to prevent further oxidation and reduction of the rxYFP protein, followed by fractionation on nondenaturing polyacrylamide gels, which were used to resolve oxidized and reduced forms of the reporter and semiquantification by an immunoblotting technique. The rxYFP was also used to determine the in vivo redox potential of the cytosolic glutathione pool in yeast, which was estimated to be highly reducing at −289 mV, indicating a very low concentration of GSSG.

**3.5. Choice of System for Leaves**

Considering the differential properties of rxYFP and roGFP discussed previously, the development of a redox reporter system in a layered multicellular organ, such as intact green leaves, favors roGFP over rxYFP for a number of reasons. Despite the advantage of having a midpoint redox potential in equilibrium with the GSH redox couple, rxYFP fluorescence emission could result from changes in halide concentrations and pH *(24)*, as well as reflecting the redox state of the targeted environment. With the use of rxYFP, it would be necessary to use a variety of controls to deconvolute the emission spectra to separate cellular redox signals from those reflected changes in halide ion and pH. In addition, rxYFP, with a single excitation maximum, prevents the use of ratiometric methods, a desirable approach in complicated tissues that eliminates the need to separately determine protein concentrations.

These limitations of rxYFPs point towards the use of a roGFP-based reporter system for application in leaves in vivo. Some of the initial problems associated with roGFPs have been overcome with the development of a pH-insensitive variant of roGFP1, roGFP1-R12, which has a less negative midpoint potential of −265 mV (approaching the midpoint potential of glutathione) and an increased response time *(21)*.

The roGFP1 variant has been successfully reported in *Arabidopsis* roots to determine the average resting redox potentials of root cytoplasm and mitochondria. Monitoring real time dynamic changes in redox in vivo, the authors demonstrated that the mitochondria are better at buffering redox changes compared with the cytoplasm *(12)*. However, in *Arabidopsis thaliana*, Haseloff and co-workers *(18)* showed that expression of GFP cDNA was curtailed by aberrant mRNA splicing. These researchers altered the codon usage of GFP to avoid recognition of a cryptic intron, which resulted in restored expression of the fluorescent protein. To develop a redox state reporter system targeting specific cellular compartments within *Arabidopsis* leaves, a similar modification would be necessary to prevent any loss of fluorescence due to splicing of the transcript by a cryptic intron.

The main advantage of using a roGFP-based system over rxYFP in leaves is that the two excitation maxima allow for the use of a ratiometric approach *(2, 12, 20)*, eliminating the need to determine protein concentrations *(see* above). A possible pitfall of using roGFPs in leaves is the fact that blue excitation wavelengths overlap considerably with the absorption spectrum of leaf chlorophylls *(25)*. Using a single excitation energy of 470 nm led to interference of chlorophyll concentration with GFP fluorescence, disrupting the proportional relationship between GFP content and fluorescence that is intrinsic to its use as a quantitative reporter *(25)*. This complication is more prevalent in plant species that show substantial changes in chlorophyll concentrations with leaf age and was less important in *Arabidopsis* where changes in chlorophyll with leaf maturation were less pronounced *(25)*.

### 3.6. Problems of Light Absorption

The relationship between chlorophyll content and roGFP signal reiterates the advantages of using a ratiometric approach to assess GFP status. However, in multilayered structures such as whole leaves, complications using duel excitation wavelengths can still be envisaged. The redox status of roGFP1 is determined with the use of a ratio of emission at 510 nm after excitation at 410 and 474 nm. However, these different excitation wavelengths are likely to be differentially absorbed dependent upon pigment concentrations, leaf thickness, and anatomy, resulting in different pathlengths and, therefore, differential penetration depths in the leaf.

It is, however, possible to establish wavelengths absorption profiles *(26)*, which would allow determination of pathlengths for excitation wavelengths and possible correction factors. It is worth noting that *Arabidopsis* leaves are significantly thinner (236 μm) compared with many other species, thus possibly shortening the pathlength in these plants *(27)*. However, it should be recognised that numerous environmental cues and stresses can cause changes in leaf pigment concentrations and composition. Modification of pigments with stress could result in different pathlengths and absorption profiles between control and treated plants. A decrease in the chlorophyll content in conjunction with an increase in xanthophyll pigments has been observed in *Pinus halepensis* needles treated with ozone, whereas in the same study, drought was shown to also increase the levels of glutathione reductase *(28)*. Pigment adjustment is also found in many mutants, for example temperature-sensitive *Arabidopsis* mutants *(29)* and UV-B tolerant *Arabidopsis (30)*.

Another possible solution to overcome differences in excitation pathlengths is to simultaneously excite and measure fluorescence emission from both the abaxial and adaxial surfaces of leaves, with appropriate excitation intensities and band-pass filters for detection. Appropriate control plants would be necessary to determine the influence of excitation pathlength on emission signals. The emission of roGFP fluorescence at 509 nm carries fewer complications and represents an advantage of roGFP over other colored variants. Green light has been shown to have one of the greatest pathlengths through the leaf *(26)* and is therefore less likely to be reabsorbed by leaf pigments as it is emitted. Emission from coloured variants of GFP, such as the blue (emission 440 nm) and cyan (emission 477 nm) will be heavily re-absorbed by leaf pigments. The red variant of GFP with an emission wavelength at 683 nm overlaps significantly with that of chlorophyll *a* fluorescence, the intensity of which would greatly exceed that emitted from any GFP.

*3.7. GFP Fluorescence Monitors*

Although historically GFP fluorescence has been detected with confocal microscopy, which requires some sort of tissue preparation, commercial systems are on the market with the capability of determining GFP signals from intact leaves. Hand-held fiberoptic fluorometers that clip onto the leaf are available (Opti-Sciences, Hudson, NH), which support a variety of source and detector combinations permitting the detection and measurement of several fluorescence markers, including GFP. Imaging systems are also available that can detect whole plant or leaf GFP signals (Qubit systems, Canada) permitting heterogeneity to be ascertained. Blue LEDs are used to provide the excitation at 490 nm whereas 510-nm emission filters in front of a CCD camera allow a GFP image to be achieved. A variety of lighting systems and optical emission filters are available for use with the other variants of GFP.

Although the commercial systems described in this chapter are for single GFP emission detection, it is possible to envisage the development of either a fibre-optic or imaging system encompassing dual excitation with detection of two emission wavelengths, which would allow roGFP detection within intact leaves. However, such a system would necessitate addressing the problems regarding differential absorption and differing pathlengths first.

## 4. Conclusions

In this chapter, we established the possibilities of using roGFPs as real-time redox reporters in soil-grown plants. With the use of specific cell promoters, it is possible to target distinct cell compartments and tissues and monitor changes in GSH redox state to determine the effects of ROS on specific cellular components. This would remove the need for confocal microscopy, which would limit the size and type of plants that could be used (13). The roGFP1-R12 is a good starting point to develop an intact plant system, as it encompasses several advantages over other variants of GFP, namely dual excitation peaks allowing ratiometric measurements, insensitivity to pH and halide ions, increased response times enabling real time measurements, and appropriate emission wavelengths allowing detection in green leaves. With the addition of an alteration to prevent mRNA splicing in *Arabidopsis*, roGFP1-R12 is an ideal candidate for developing a redox reporter system in whole plants.

## References

1. Apel, K. and Hirt, H. (2004) Reactive oxygen species: metabolism, oxidative stress, and signal transduction. *Ann Rev Plant Biol* **55**, 373–399.

2. Hanson, G. T., Aggelers, R., Oglesbees, D., Cannon, M., Capaldis, R. A., Tsien, R. Y., and Remington, S. J. (2004) Investigating mitochondrial redox potential with redox-sensitive green fluorescent protein indicators. *J Biol Chem* **279**, 13044–13053.

3. LeMoan, N., Clement, G., Le Maout, S., Tacnet, F., and Toledano, M. B. (2006) The *Saccharomyces cerevisiae* proteome of oxidized protein thiols. *J Biol Chem* **281**, 10420–10430.

4. Mullineaux, P. M., Karpinski, S., and Baker, N. R. (2006) Spatial dependence for hydrogen peroxide-directed signalling in light-stressed plants. *Plant Physiol* **141**, 346–350.

5. Bechtold, U., Karpinski, S., and Mullineaux, P. M. (2005) The influence of the light environment and photosynthesis on oxidative signalling responses in plant-biotrophic pathogen interactions. *Plant Cell Environ* **28**, 1046–1055.

6. Gapper, C. and Dolan, L. (2006) Control of plant development by reactive oxygen species. *Plant Physiol* **141**, 341–345.

7. Torres, M. A., Jones, J. D.G., and Dangl, J. L. (2006) Reactive oxygen species signalling in response to pathogens. *Plant Physiol* **141**, 373–378.

8. Fryer, M. J., Ball, L., Oxborough, K., Karpinski, S., Mullineaux, P. M., and Baker, N. R. (2003) Control of *Ascorbate Peroxidase 2* expression by hydrogen peroxide and leaf water status during excess light stress reveals a functional organisation of *Arabidopsis* leaves. *Plant J* **33**, 691–705.

9 Ushio-Fukai, M. (2006) Localizing NADPH oxidase-derived ROS. *Science STKE*. www.stke.org/cgi/content/full/2006/349/re8

10. Mullineaux, P. M. and Rausch, T. (2005) Glutathione, photosynthesis and the redox regulation of stress-responsive gene expression. *Photos Res* **86**, 459–474.

11. Noctor, G., Gomez, L., Vanacker, H., and Foyer, C. H. (2002) Interactions between biosynthesis, compartmentation and transport in the control of glutathione homeostasis and signalling. *J Exp Bot* **53**, 1283–1304.

12. Jaing, K., Schwarzer, C., Lally, E., Zhang, S., Ruzin, S., Machen, S., Remington, J., and Feldman, L. (2006) Expression and characterization of a redox-sensing green fluorescent protein (reduction-oxidation-sensitive green fluorescence protein) in *Arabidopsis*. *Plant Physiol* **141**, 397–403.

13. Fricker, M., Runions, J., and Moore, I. (2006) Quantitative fluorescence microscopy: From art to science. *Ann Rev Plant Biol* **57**, 79–109.

14. Hartmann, T. N., Fricker, M. D., Rennenberg, H., and Meyer, A. J. (2003) Cell-specific measurement of cytosolic glutathione in poplar leaves. *Plant Cell Environ* **26**, 965–975.

15. Gutierrez-Alcala, G., Gotor, C., Meyer, A. J., Fricker, M., Vega, J. M., and Romero, C. (2000) Glutathione biosynthesis in *Arabidopsis* trichome cells. *Proc Natl Acad Sci USA* **97**, 11108–11113.

16. Chalfie, M. and Kain, S. (2005) *Green Fluorescence Protein. Properties, Applications and Protocols. Methods of Biochemical Analysis.* Wiley, London/New York.

17. Ostergaard, H., Henriksen, A., Hansen, F. G., and Winther, J. R. (2001) Shedding light on disulfide bond formation: engineering a redox switch in green fluorescent protein. *EMBO J* **20**, 5853–5862.

18. Haseloff, J., Siemering, K. R., Prasher, D. C., and Hodge, S. (1997) Removal of a cryptic intron and subcellular localization of green fluorescence protein are required to mark transgenic *Arabidopsis* plants brightly. *Proc Natl Acad Sci USA* **94**, 2122–2127.

19. Ostergaard, H., Tachibana, C., and Winther, J. R. (2004) Monitoring disulfide bond formation in the eukaryotic cytosol. *J Cell Biol* **166**, 337–345.

20. Dooley, C. T., Dore, T. M., Hanson, G. T., Jackson, W. C., Remington, S. J., and Tsien, R. Y. (2004) *J Biol Chem* **279**, 22284–22293.

21. Canon, M. B. and Remington, S. J. (2006) Re-engineering redox-sensitive green fluorescence protein for improved response rate. *Protein Sci* **15**, 45–57.

22. Hanson, R. E., Ostergaard, H., and Winther, J. R. (2005) Increasing the reactivity of an artificial dithiol-disulfide pair through modification of the electrostatic milieu. *Biochemistry* **41**, 5899–5906.

23. Remington, S. J. (2000) Structural basis for understanding spectral variations in green fluorescence protein. Bioluminescence and chemiluminescence, *Methods Enzymol* **305**, 196–211.

24. Griesbeck, O., Baird, G. S., Campbell, R. E., Zacharias, D. A., and Tsien, R. Y. (2001) Reducing the environmental sensitivity of yellow fluorescent protein. *J Biol Chem* **276**, 29188–29194.

25. Zhou, X., Carranco, R., Vitha, S., and Hall, T. C. (2005) The dark side of green fluorescent protein. *New Phytol* **168**, 313–322.

26. Vogelmann, T. V. and Evans, J. R. (2002) Profiles of light absorption and chlorophyll within spinach leaves from chlorophyll fluorescence. *Plant Cell Environ* **25**, 1313–1323.

27. Ramonell, K. M., Kuang, A., Porterfield, D. M., Crispi, M. L., Xiao, Y., McClure, G., and Musgrave, M. E. (2001) Influence of atmospheric oxygen on leaf structure and starch deposition in *Arabidopsis thaliana*. *Plant Cell Environ* **24**, 419–428.

28. Alonso, R., Elvira, S., Castillo, J., and Gimeno, B. S. (2001) Interactive effects of ozone and drought stress on pigments and activities of antioxidative enzymes in *Pinus halepensis*. *Plant Cell Environ* **24**, 905–916.

29. Araki, N., Kusumi, K., Masamoto, K., Niwa, Y., and Iba, K. (2000) Temperature-sensitive Arabidopsis mutant deficient in 1-deoxy-D-xylulose-5-phosphate synthase within the plastid non-mevalonate pathway of isoprenoid biosynthesis. *Physiol. Plantarum.* **108**, 19–24.

30. Bieza, K. and Lois, R. (2001) An Arabidopsis mutant tolerant to lethal ultraviolet-B levels shows constitutively elevated accumulation of flavinoids and other phenolics. *Plant Physiol* **126**, 1105–1111.

# Chapter 6

# Imaging of Intracellular Hydrogen Peroxide Production With HyPer Upon Stimulation of HeLa Cells With Epidermal Growth Factor

## Kseniya N. Markvicheva, Ekaterina A. Bogdanova, Dmitry B. Staroverov, Sergei Lukyanov, and Vsevolod V. Belousov

## Abstract

Reactive oxygen species (ROS) regulate both normal cell functions by activating a number of enzymatic cascades and pathological processes in many diseases by inducing oxidative stress. For many years since the discovery of ROS in biological systems, there were no adequate methods of detection and quantification of these molecules inside the living cells. We developed the first genetically encoded fluorescent indicator for the intracellular detection of hydrogen peroxide, HyPer, that can be used for imaging of $H_2O_2$ production by cells under various physiological and pathological conditions. Unlike most known ROS indicators, HyPer allows the generation of a real-time image series that give precise information about the time course and intensity of $H_2O_2$ changes in any compartment of interest. In this chapter, we describe the method of confocal imaging of hydrogen peroxide production in HeLa cells upon stimulation with epidermal growth factor. The technique described may be accepted with minimal variations for the use in other cell lines upon various conditions leading to $H_2O_2$ production.

**Keywords**: Hydrogen peroxide, EGF, fluorescent indicator, fluorescent protein, YFP, HyPer, ROS, confocal microscopy.

## 1. Introduction

Since the discovery of reactive oxygen species (ROS) and their role in cell physiology and pathology (1), a number of accurate techniques for the measurement of ROS in vitro have

John T. Hancock (ed.), *Methods in Molecular Biology, Redox-Mediated Signal Transduction, vol. 476*
© 2008 Humana Press, a part of Springer Science+Business Media, Totowa, NJ
DOI: 10.1007/978-1-59745-129-1_6

been developed and are now widely used. Most of them can be applied to model systems such as isolated proteins, organelles, cell or tissue extracts, or permeabilized cells. However, most attempts to directly measure ROS in intact cells were not very successful. Often, information about ROS participation in biochemical processes is mostly based on indirect evidence, for example, the level of DNA/lipids/proteins oxidation. Also, the significance of the data obtained with optical fluorescent probes for ROS detection is now not clear because of the high risk of artifacts. The most known dichlorofluorescein (DCF) derivatives fluoresce after oxidation by ROS. However, these dyes could be oxidized by different types of ROS and reactive nitrogen species (2). Moreover, DCF derivatives are able to produce ROS upon exposure to light (3, 4), which leads to an artifactual ROS generation and signal amplification. Like most chemical probes, these dyes cannot be targeted to various intracellular compartments known to be the major ROS sources–such as plasma membrane and mitochondria.

Recently, we developed the first genetically encoded fluorescent indicator for intracellular hydrogen peroxide detection, named HyPer (5). It consists of circularly permuted yellow fluorescent protein (cpYFP) inserted into the regulatory domain of the *Escherichia coli* $H_2O_2$-sensing protein, OxyR. OxyR contains an $H_2O_2$-sensitive regulatory domain (amino acids 80–310) and a DNA-binding domain (amino acids 1–79). Upon oxidation by $H_2O_2$, the reduced form of OxyR is converted into an oxidized DNA-binding form. The key residues of OxyR in this regard are Cys199 and Cys208 (6). Cys199 resides within a hydrophobic pocket, and exposure of OxyR to $H_2O_2$ converts Cys199 to a charged intermediate that is released from the hydrophobic surround and forms a disulfide bridge with Cys208. As a result, a dramatic conformational change occurs in the regulatory domain of OxyR, especially within a flexible region located at residues 205–222 (7).

We produced HyPer by inserting cpYFP into the regulatory domain of OxyR between residues 205 and 206. HyPer demonstrates a submicromolar affinity to $H_2O_2$, and at the same time it is insensitive to other oxidants. HyPer has two excitation peaks at 420 and 500 nm and one emission peak at 516 nm. Upon exposure to $H_2O_2$, the excitation peak at 420 nm decreases proportionally to the increase in the peak at 500 nm in a ratiometric manner.

In contrast to DCF derivatives, HyPer is genetically encoded and therefore could be targeted to various subcellular compartments. HyPer is able to detect relatively high concentrations of $H_2O_2$ produced during apoptosis as well as low-level $H_2O_2$ produced in cells upon physiological stimulation by growth factors. HyPer can be rapidly oxidized by submicromolar $H_2O_2$ even in a highly reduced intracellular environment. Thus, HyPer lacks disadvantages of chemical probes for ROS detection. The use of HyPer significantly simplifies both monitoring of changes in $H_2O_2$ levels and in the interpretation of the results.

In this chapter, we describe a protocol for fluorescent confocal HyPer imaging of changes in $H_2O_2$ level in the cytoplasm of HeLa cells stimulated with epidermal growth factor (EGF). The protocol is short and simple and can be applied with minimal variations for other cell lines upon different conditions where a change in the level of $H_2O_2$ inside the cell is suspected.

# 2. Materials

### 2.1. Cell Culture and Transfection

1. Dulbecco's Modified Eagle Medium (D-MEM; Invitrogen) supplemented with 10% fetal calf serum (FCS; PAA Laboratories).

2. Opti-MEM without phenol red (Invitrogen) and without serum.

3. FCS (PAA Laboratories).

4. FuGENE 6 transfection reagent (Roche).

5. GIBCO™ Leibovitz's L-15 Medium (Invitrogen).

6. Thirty-five millimeter glass-bottom dishes (World Precision Instruments, http://www.wpiinc.com)

7. HyPer-Cyto vector for mammalian expression (Evrogen, http://www.evrogen.com). The vector is usually supplied at concentration 500 ng/µL.

8. Human EGF (Sigma) is dissolved at 2 mg/mL in DMEM and stored in single-use aliquots at –20°C. A working solution (100 µg/mL) is prepared by dilution in 100 µg/mL bovine serum albumin.

# 3. Methods

### 3.1. Transfection

1. Seed HeLa cells onto 35-mm glass-bottom dishes in D-MEM supplemented with 10% FCS.

2. Allow the cells to grow to 70–80% confluence.

3. Change media for Opti-MEM without serum, 700 µL per dish. Incubate 1 h. **Steps 4–7** describe the protocol for one 35-mm glass-bottom dish:

4. To a sterile eppendorf tube (1.5 ml) add 100 µl serum free Opti-MEM. Add 3 µl of FuGENE 6 transfection reagent (*see* **Note** 1). Mix rapidly by vortexing, and allow to stay for 5 min.

5. Add 2 µl (1 µg) of HyPer-Cyto vector. Mix rapidly by vortexing, allow to stay for 15 min for liposome-DNA complex to form.

6. Add the content of the tube to the media covering the cells in the glass bottom dish. Mix gently.

7. Add FCS to reach final concentration of 2%. Mix gently. Place the dish into a $CO_2$ incubator.

### 3.2. Imaging of $H_2O_2$ Induction in HeLa Cells Stimulated With EGF

1. Sixteen to 24 h after transfection, check the effectiveness of transfection by using excitation with a mercury lamp. Use the GFP filter set in "visual" mode of the microscope. In fact, it is not necessary to get 100% of the cells transfected, but the more fluorescent cells that are in a field of view, the more information about the synchronicity of the cells' response during imaging series is obtained.

2. Experiments can be carried out without a $CO_2$ and temperature control chamber. One hour before imaging, change Opti-MEM supplemented with 2% FCS for 1 mL of L15 media supplemented with 2% FCS and allowed the dish to stay at room temperature.

3. Place the dish on the microscope table (*see* **Note** 2), and find a group of transfected cells by using the visual mode of the microscope. Once the cells are in the field of view, immediately close the lamp shutter to avoid unwanted bleaching of the probe.

4. Switch the microscope to the scanning mode and activate the following beam path settings, or equivalent depending on the manufacture of the microscope:
   a. "Mode"—xyt. This mode is used for acquisition of time series for two-dimensional images (*see* **Note** 3).
   b. "Format"—512 × 512 (*see* **Note** 4).
   c. "Beam exp."—3.
   d. Pinhole—2 airy.
   e. Scan direction—unidirectional.
   f. PMT voltage—730–800. Lower value makes pictures less noisy; higher value lets the use of less laser power.
   g. Ar laser power (488 nm line)—5–15%.
   h. Excitation Beam Splitter DD 488/543.
   i. Emission wavelengths—500–530 nm.
   J. "Speed" (Beam frequency) 400 Hz.
   If available; optionally, "line average 2" function can be activated to make pictures less noisy.

Set number of frames to be acquired and delay between frames using "Time" function. For growth factors stimulation experiments usually the use 1–10 s delay is recommended. All imaging experiments usually take 1–2 h.

Zoom selected cells with "Zoom in" function.

5. Perform several single scans to set final focus; adjust laser power and PMT voltage. Press "Start Series" button. The YFP that is the fluorescent core of HyPer tends to photoconvert into the dark state upon irradiation with excitation light. After several seconds in the dark state, YFP chromophore turns back into the fluorescent state. At every single moment, there is a dynamic equilibrium between fluorescent and dark states that depends on excitation light power and frequency of frames acquisition. In fact, during the several frames after the start of a time series, fluorescence goes down and after the equilibrium between fluorescent and dark states of YFP is stable fluorescence no longer changes. Press "Stop Series" button and than immediately press "Start Series" button again. During this time the next 10–20 frames will give a "baseline". Acquiring a series can be then achieved by activating "Quantification–stack profile" window of Leica Confocal Software (LCS) software.

6. Dilute 1 µL of working solution of EGF in 50 µL of L-15 media and carefully, drop-by-drop add it to the dish. Spread single drops to "cover" all surface. Do not touch the dish during addition. If you use the liquid perfusion module (*see* **Note** 5), perfuse L-15 media with the same concentration of EGF. Mark the time point of addition.

7. Stop the acquisition at the appropriate time.

**3.3. Quantification of Results**

Quantification can be easily performed with the standard LCS software.

1. Press the "LUT" button. Select "P.color 7" lookup table (*see* **Note** 6).

2. Activate "Process" window in LSC software, select "Enhancement" window, then choose the "Baseline correction" option to allow subtract background. After background subtraction, fluorescent signal from the cells can be quantified.

3. Open "Quantify" window of LCS software (**Fig.** 6.1). Activate "Stack profile" window. Press "Viewer" button.

4. Select regions of interest.

5. Press "Graphs" button. In the window that appears the time profile of fluorescence intensity in selected regions of interest will be seen. Activation of the right mouse button menu on the graphs allows the exporting of graphs to ".txt" or ".jpeg" formats (*see* **Note** 7).

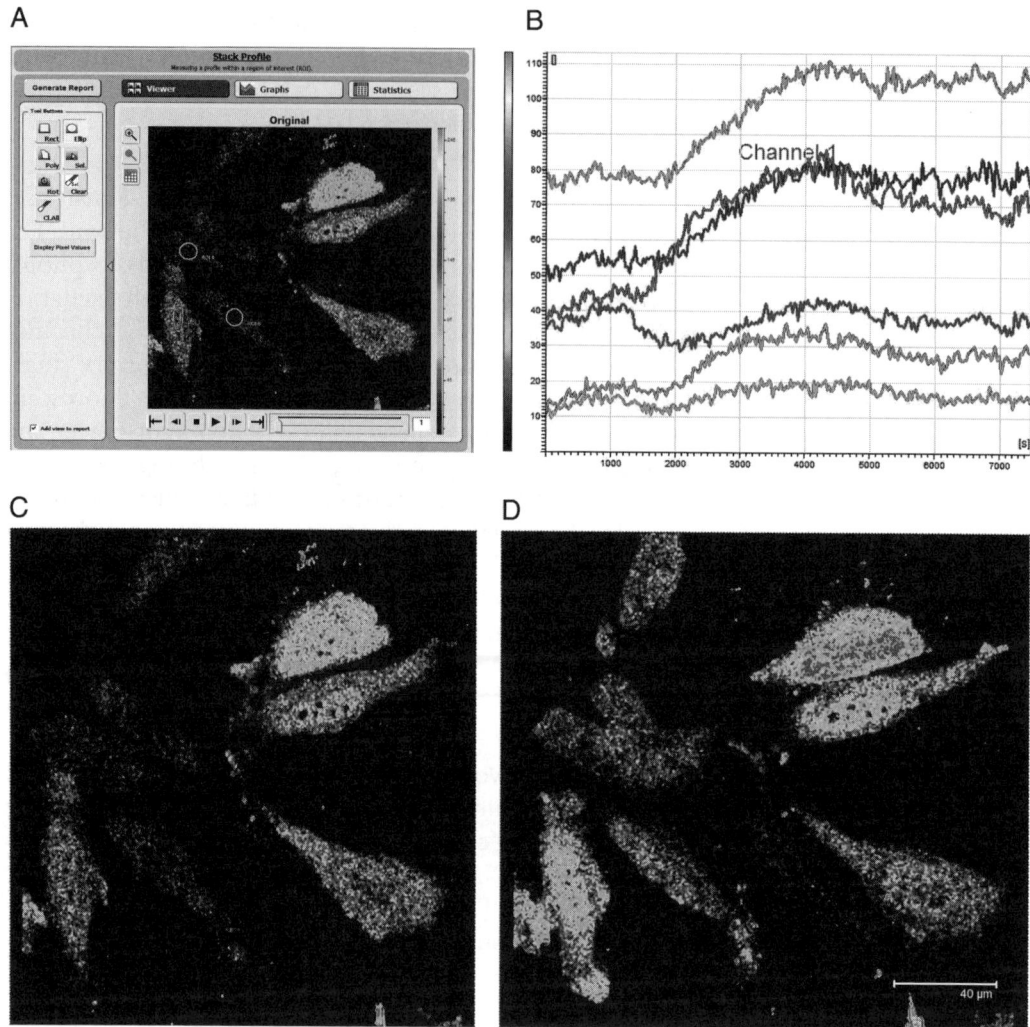

Fig. 6.1. Quantification of HyPer imaging series. **(A)** Screen shot from Leica Confocal Software "Stack Profile—Viewer" window. In this window all the stack can be viewed. The left upper buttons allow differently shaped regions of interest to be defined. By clicking the "Graphs" button above the image, time course **(B)** of signal intensities in selected regions of interest appears. Activating the right mouse button menu on the graphs allows the export of graphs to ".txt" or ".jpeg" formats. Pseudo-colored images of HeLa cells expressing HyPer-Cyto at the time point of **(C)** and 1 h after **(D)** EGF addition are shown (*See* Color Plates).

## 4. Notes

1. HeLa cells are easily transfected by most transfection reagents. It is our preference to use FuGENE 6 if given because of its simple protocol of use and low cytotoxicity. Following

the recommendations from the developer, use Opti-MEM media without serum for transfection with FuGENE 6. The main recommendation from the developer of FuGENE 6 is that contact with plastic surfaces other than pipet tips should be avoided. Add FuGENE 6 directly into the volume of the Opti-MEM media.

2. For imaging, if available, use Leica confocal system TCS SP2 on inverted microscope Leica DM IRE equipped with HCX PL APO lbd.BL 63x 1.4NA oil objective and 125-mW Ar and 1-mW HeNe lasers.

3. If cells used do not significantly change their shape during the course of imaging, the xt (line scan) mode can be used as it is less photo-damaging for the cells.

4. The lower resolution gives less detailed pictures, and the higher resolution slows the laser beam, which is more photo-damaging for the cell. In any experiment with the real-time dynamics of fluorescence, two applications are possible: to quantify the fluorescent signal or to make "pretty" pictures. For quantification, the use of low resolution is enough and much less damaging.

5. There are many liquid perfusion modules on the market. If the microscope one is using is equipped with one of them, add the compounds according to the instruction of module manufacturer. Also, if the microscope being used is equipped with $CO_2$ control module, use Opti-MEM instead of L-15 media.

6. Differences in brightness between frames are better represented not in monochrome but by using a multicolor Lookup Table.

7. Leica format ".lei" is a stack of TIFF files. These files could be quantified by any analysis software such as freeware "ImageJ" (Waine Rasband, NIH, http://rsb.info.nih.gov/ij/).

# Acknowledgments

This work was supported by grants from the European Commission (FP-6 Integrated Project LSHG-CT-2003-503259), the Russian Academy of Sciences Program in Molecular and Cell Biology, the Russian Foundation for Basic Research (Project 07-04-12189), the National Institutes of Health (GM070358), and the Howard Hughes Medical Institute grant HHMI 55005618.

## References

1. Droge, W. (2002) Free radicals in the physiological control of cell function. *Physiol Rev* **82**, 47–95.

2. Crow, J. P. (1997) Dichlorodihydrofluorescein and dihydrorhodamine 123 are sensitive indicators of peroxynitrite in vitro: implications for intracellular measurement of reactive nitrogen and oxygen species. *Nitric Oxide* **1**, 145–157.

3. Marchesi, E., Rota, C., Fann, Y. C., Chignell, C. F., and Mason, R. P. (1999) Photoreduction of the fluorescent dye 2′-7′-dichlorofluorescein: a spin trapping and direct electron spin resonance study with implications for oxidative stress measurements. *Free Radic Biol Med* **26**, 148–161.

4. Rota, C., Fann, Y. C., and Mason, R. P. (1999) Phenoxyl free radical formation during the oxidation of the fluorescent dye 2′,7′-dichlorofluorescein by horseradish peroxidase. Possible consequences for oxidative stress measurements. *J Biol Chem* **274**, 28161–28168.

5. Belousov, V. V., Fradkov, A. F., Lukyanov, K. A., Staroverov, D. B., Shakhbazov, K. S., Terskikh, A. V., and Lukyanov, S. (2006) Genetically encoded fluorescent indicator for intracellular hydrogen peroxide. *Nat Method* **3**, 281–286.

6. Zheng, M., Aslund, F., and Storz, G. (1998) Activation of the OxyR transcription factor by reversible disulfide bond formation. *Science* **279**, 1718–1721.

7. Choi, H., Kim, S., Mukhopadhyay, P., Cho, S., Woo, J., Storz, G., and Ryu, S. (2001) Structural basis of the redox switch in the OxyR transcription factor. *Cell* **105**, 103–113.

# Chapter 7

## Tools to Investigate Reaction Oxygen Species-Sensitive Signaling Proteins

### Radhika Desikan, Steven J. Neill, Jenna Slinn, and John T. Hancock

### Abstract

The thiol groups of cysteine residues on proteins are attractive oxidative targets for modification by reactive oxygen species, such as hydrogen peroxide ($H_2O_2$). Such modification can lead to important cellular signaling processes that ultimately result in modification of the physiology of the organism. To identify such proteins that are amenable to oxidative modification, different methods can be used. Here, two such approaches are described: one being the use of fluorescent thiol derivatives, and the second being the use of genetic mutants that are mutated in thiol residues. Using the model plant *Arabidopsis thaliana*, cell cultures, and whole plants, we describe these tools to help the reader understand the function of such thiol modifications on plant responses.

**Keywords:** Glyceraldehyde 3-phosphate dehydrogenase, histidine kinase, hydrogen peroxide, 5′-iodoacetamide fluorescein, reactive oxygen species, stomatal guard cells, thiol labeling

## 1. Introduction

Modification of thiol residues on proteins via redox signals is an important mechanism by which redox signal transduction occurs in cells. Sensing of redox signals by these residues cause changes in the structure and hence function of the protein, thereby initiating a series of signaling cascades *(1)*. Identification of such proteins is therefore key towards understanding how redox signaling occurs. To identify such proteins, different approaches can be used, such

John T. Hancock (ed.), *Methods in Molecular Biology, Redox-Mediated Signal Transduction, vol. 476*
© 2008 Humana Press, a part of Springer Science+Business Media, Totowa, NJ
DOI: 10.1007/978-1-59745-129-1_7

as tagging thiol residues with detectable groups such that the tagged protein can be subsequently identified or using proteins mutated in thiol residues, either as a pure protein or in its native organism, to analyze its function. Examples of these will be presented in this chapter.

### 1.1. Outline to the 5'-iodoacetamide fluorescein (IAF) Procedure

Fluorescent covalently modified iodoacetamide can react with free thiol groups on proteins, such that when the protein is preoxidized by reactive oxygen species (ROS), the iodoacetamide group cannot react with the thiol group. Therefore, we can assess the effect of ROS on thiol modification by conducting experiments in the presence and absence of ROS, along with the iodoacetamide dye, subsequently followed by protein analysis. Using such an approach, it has recently been shown that the plant enzyme cytosolic glyceradehyde 3-phosphate dehydrogenase (GAPDH) is a target of ROS modification *(2)*. In the first tool described in this chapter, the iodoacetamide labeling procedure followed by proteomics analysis of the labeled proteins is described. It is important to note that the concentrations of ROS that are used in these experiments are physiological to the system that is being used here, which in this case are cell cultures of *Arabidopsis thaliana*.

### 1.2. Outline to the Use of Mutants

The advantage of using an organism whose genome has been sequenced, such as *Homo sapiens* or *A. thaliana*, is that a number of mutants are readily available. A huge international community has contributed to a wealth of knowledge in the model plant *A. thaliana*, which has also led to development of various public resources, such as the availability of seed stocks of different mutants via stock centers. Using this approach, a family of genes called histidine kinases were screened to test their ability to act as ROS sensors. One group of proteins in this family belongs to the class of ethylene receptors. Ethylene is a plant hormone involved in a number of physiological processes, such as fruit ripening, the falling of leaves, and the formation of root hairs. A group of five receptors of the histidine kinase family act as ethylene receptors. A large amount of research into ethylene signaling has involved the use of *Arabidopsis* plants mutated in these receptors. One such mutant is mutated in a Cys residue at the ethylene binding site. Therefore, the role of this Cys, and hence this protein, in ROS signaling in plants has been explored by studying different physiological responses to ROS *(3)*. Some of these include the effect of hydrogen peroxide on stomatal closure, gene expression and root growth. By comparison of wild-type and mutant plants grown under exactly the same conditions, analysis of the requirement for a functional protein in ROS signalling can be achieved.

## 2. Materials

1. AT3 medium: Murashige and Skoog medium supplemented with 3% sucrose, 0.5 mg/L NAA and 0.05 mg/L kinetin, pH 5.5.

2. Murashige and Skoog media (Sigma, UK, cat. no. M5515).

3. Extraction buffer: 50 mM Tris-HCl, pH 7.5, with complete protease inhibitor cocktail.

4. Equilibration solution: 50 m$M$ Tris-HCl, pH 8.8; 6 $M$ urea; 30% glycerol; 2% sodium dodecyl sulfate (SDS); 0.025% bromophenol blue.

5. Protease inhibitor cocktail tablets (Roche).

6. 5'-iodoacetamide fluorescein (5'-IAF) made as a stock solution fresh (for example, 10 m$M$ stock in dimethyl sulfoxide).

7. Two-dimensional electrophoresis solutions and reagents: all PlusOne reagents (GE Healthcare, UK).

8. $H_2O_2$ solution (30% v/v; Sigma-Aldrich Co., Ltd., Dorset UK).

9. Stomatal bioassay buffer: 10 m$M$ MES; 5 m$M$ KCl; 100 µ$M$ CaCl$_2$, pH 6.15; with 1 $M$ Tris-HCl, pH 7.5.

10. Levington's F2 compost with sand.

11. Glyceraldehyde 3-phosphate dehydrogenase (GAPDH) assay buffer: 100 m$M$ Tris-HCl, pH 8; 10 m$M$ MgCl$_2$; 0.2 m$M$ NADH; 5 m$M$ ATP; 2 m$M$ phosphoglyceric acid.

12. Isolated GAPDH enzyme, for example, from rabbit muscle (Sigma).

13. MES/KCl buffer: 5 m$M$ KCl, 10 m$M$ MES, 50 µ$M$ CaCl$_2$, pH 6.15.

14. Seed sterilizing solution: 10% bleach, 0.1% SDS.

## 3. Methods

### 3.1. Growth and Maintenance of Cell Suspension Cultures

Suspension cultures of *A. thaliana* (obtained from an established source) must be maintained under aseptic conditions as follows:

1. Once a week, transfer 10 mL of a 7-d culture into 100 mL of AT3 medium in sterile conical flasks (*see* **Note** 1). This should be performed in a laminar flow cabinet.

2. Maintain the cells in a controlled growth environment room/chamber at 24°C in the light (16 h, 100–150 µE/m²/s) on a rotary shaker at 110 rpm.

3. When cell cultures are required for treatment, aliquot out the required volume of cells (note their packed cell volume) into sterile Falcon tubes or equivalent. Following treatment, harvest the cells by vacuum filtration through Whatman paper and a Buchner funnel. Cells are subsequently frozen in liquid nitrogen and stored at –80°C until further use.

### 3.2. Treatment of Cells and Labeling of Proteins

1. For treatment with $H_2O_2$, cell cultures of equal cell densities are aliquoted into sterile Falcon tubes, or equivalent. Concentrations of $H_2O_2$ that are physiological to the system being used (in this case, cell cultures) are applied to the cell cultures and left for a defined time period. For example, apply 10 m$M$ $H_2O_2$ to cell cultures of a known volume, for 10 min with constant shaking (*see* **Note** 2). Once the treatment is done, harvest the cells by filtration as described above: **step 3** of Section 3.1.

2. Extract proteins from harvested cells as follows. Grind the filtered cell cultures in a prechilled mortar and pestle at 4°C in extraction buffer (*see* **Note** 3).

3. Centrifuge the homogenate in microfuge tubes at 12,000 $g$ for 10 min.

4. Remove the supernatant and recentrifuge this supernatant in a fresh tube at 12,000 $g$ for 10 min.

5. Determine protein concentrations using a reliable protein determination assay (e.g., Bradford assay or equivalent).

6. For labeling the proteins, use equal concentrations of proteins across the treatment range, in a defined volume, i.e., adjust the volume of the samples so that they are equal. To each protein extract, add 5′-IAF to a final concentration of 100 µ$M$ (from a stock solution of 5′-IAF of 10 m$M$ in dimethyl sulfoxide) and incubate in the dark at room temperature for 10 min.

### 3.3. Analysis of Proteins With Electrophoresis

Proteins labeled as described in Section 3.2 can be analyzed by either one- or two-dimensional gel electrophoresis. For one-dimensional gel electrophoresis (1D-SDS), use standard procedures as used for SDS–polyacrylamide gel electrophoresis, using a 12% resolving gel, and in the dark to avoid photo-bleaching of the dye. Once completely resolved, scan the gels using a Typhoon scanner (GE Healthcare, Bucks, UK) at an excitation wavelength of 490–495 nm and emission wavelength of 515–520 nm. For identification of proteins, subsequently stain the gels with Colloidal Coomassie as per the manufacturer's guidelines (Sigma).

For two-dimensional gel electrophoresis (2D gel electrophoresis), different procedures are followed, which involve the accurate quantification followed by electrophoresis according to the pI and subsequently the molecular weight of the proteins being analysed. After one performs electrophoresis, labeled proteins can be visualized (Typhoon scanner [GE Healthcare] at an excitation wavelength of 490–495 nm and emission wavelength of 515–520 nm). Then, the identification of proteins can be performed with mass spectrometry. The procedures are described briefly below; however, methods differ between different manufacturers. The procedure described here is according to that prescribed by GE Healthcare, UK for 2D gel electrophoresis.

1. Extract the proteins as described in **steps 1–4** of Section 3.2.

2. Determine the concentration of soluble proteins using a 2D Quantification kit (GE Healthcare).

3. Add 100 µ$M$ 5′-IAF to the protein extracts (of equal concentration, 50 µg) for 10 min at room temperature.

4. Precipitate the extracts with trichloroacetic acid (TCA)/acetone (*see* **Note** 5) and resuspend the pellet in rehydration buffer (8 $M$ urea, 2% CHAPS).

5. Add 0.5% IPG buffer and 2.8 mg/ml DTT to the samples, and analyse by electrophoresis as follows: IEF overnight on 24 cm pH 3–10 nonlinear Immobiline DryStrip gels (GE Healthcare) using an IPGPhor unit, in the dark, with the following settings: 500 V for 10 min, 4000 V for 1 h 30 min, 8000 V for 6 h 30 min at 20°C. At this point, the strips are ready to be electrophoresed on the second dimension.

6. For the second dimension, strips are first equilibrated for 15 min in equilibration solution containing 0.5% (w/v) DTT and then for 15 min in equilibration solution containing 4.5% (w/v) iodoacetamide for 15 min.

7. Resolve the proteins by loading the strips onto 12.5% SDS gels prepared using the Ettan DALTsix gel system (GE Healthcare) at 20°C at an initial power setting of 2.5 W per gel for 30 min, followed by 18 W per gel for 5 h.

8. After electrophoresis is performed, scan the gels on the Typhoon scanner as described earlier in this subheading.

9. After scanning, stain the gels using Colloidal Coomassie stain (Sigma) according to the manufacturer's guidelines.

*3.4. Identification of Proteins With Matrix-Assisted Laser Desorption /Ionization Time-of-Flight Mass Spectrometry*

After electrophoresis, any spots or bands identified as being labelled with 5′-IAF have to be identified. For this, the protein band or spot is excised with a sharp scalpel blade from the acrylamide gel and under very clean conditions, placed in a 1.5-mL tube. The procedure is repeated for every spot or band identified as being differentially labeled on the gel. Samples are subsequently digested using trypsin.

1. Chop the spot/band into small (1 × 1 mm) pieces, this is best done in the tube using a clean scalpel or needle.

2. Wash the gel pieces in 150 µL of water for 5 min. Centrifuge for 3 min and remove supernatant.

3. Add 150 µL of acetonitrile and leave for 10–15 min to dehydrate the gel pieces.

4. Remove all acetonitrile and add approximately 50 µL of 10 m$M$ DTT in 0.1 m$M$ NH$_4$HCO$_3$. Incubate for 30 min at 56°C.

5. Remove DTT and add acetonitrile to dehydrate the gel pieces as before.

6. Centrifuge for 3 min and remove supernatant and add 50 µL of 55 m$M$ iodoacetamide in 0.1 m$M$ NH$_4$HCO$_3$ and leave in the dark for 45 min.

7. Centrifuge for 3 min and remove supernatant and wash with 150 µL of 0.1 m$M$ NH$_4$HCO$_3$ for 15 min.

8. Centrifuge for 3 min and remove supernatant and add acetonitrile to dehydrate the gel pieces as before.

9. Dry the gel pieces in a vacuum centrifuge.

10. Prepare a trypsin solution in 50 m$M$ NH$_4$HCO$_3$ containing 12.5 ng/µL trypsin.

11. Add 20 µL of this trypsin solution and incubate at 4°C for 30–45 min. Check after 15–20 min and top up with more trypsin solution if all liquid has been absorbed by the gel piece.

12. After 30–45 min, remove all remaining liquid and add 20 µL of 50 m$M$ NH$_4$HCO$_3$ without trypsin.

13. Incubate at 37°C for 16 h.

14. To extract the peptides, add 10 µL of 50% acetonitrile/5% formic acid and place tube in a sonicating water bath for 10 min.

15. Remove liquid and transfer to a clean tube (this should contain the peptides).

16. Repeat the sonication once more using 10–20 µL of 50% acetonitrile/5% formic acid, each time transferring the liquid to the clean tube.

17. Mix an aliquot of the extracted peptides (5 µL) with 5 µL of matrix solution (0.05 $M$ α-cyano-4-hydroxycinnamic acid in methanol/acetonitrile [1:1]).

18. Analyse an aliquot of the sample (1 µL) with matrix-assisted laser desorption/ionizationtime-of-flight mass spectrometry mass spectrometry (Waters-Micromass, Manchester, UK).

19. To confirm the identity of proteins, peptides can also be analysed with LC-electrospray tandem mass spectrometry (Q-tof micro, Waters-Micromass), using a C18 Pepmap

column, with a 5–80% acetonitrile gradient with 0.1% formic acid, over the course of 60 min with a flow rate of 200 nL/min. Perform electrospray tandem mass spectrometry in data-dependent acquisition mode with a capillary voltage of 3.5 kV (*see* **Note 4**).

20. Bioinformatic analyses should be carried out using Protein-Lynx Global Server (Waters Micromass).

With the use of the aforementioned procedures, we identified GAPDH (accession number P25858), from *Arabidopsis* suspension cultures, as being labeled with 5′-IAF, and in the presence of $H_2O_2$ the fluorescence intensity of this protein decreased significantly. On the basis of these findings, we concluded that GAPDH in *Arabidopsis* cell cultures is a target of $H_2O_2$ signaling *(2)*. To prove a functional role for this, it is important to ascertain whether or not redox modification would actually change the function of the protein that has been labelled. To test this, assays for the activity of GAPDH have to be performed, and the response to $H_2O_2$ tested.

### 3.5. Assays for GAPDH Activity

To determine whether $H_2O_2$ does affect the function of the protein GAPDH identified previously using thiol-tagging and proteomics, activity of the enzyme GAPDH is tested in the absence and presence of $H_2O_2$. For this either pure commercial enzyme or plant extract (soluble proteins, as obtained in **steps 1–4** of Sect. 3.2) can be used.

GAPDH activity can be determined using a coupled NADH-dependent reaction. The reactions catalysed are as follows:

Equation 1:

$$PGA + ATP \xrightarrow{\text{(PPK, } MgCl_2)} 1,3 - \text{bisphosphoglycerate} + ADP$$

Equation 2:

$$1,3 - \text{bisphosphoglycerate} + NADH + H^+ \xrightarrow{\text{(GAPDH)}}$$

$$\text{glyceraldehyde } 3 - \text{phosphate} + NAD^+ + P_i$$

1. Into a cuvette, place 3 mL of GAPDH assay buffer (*see* Sect. 2).

2. Add either 35 units of the isolated enzyme or soluble protein extract from plants (*see* **Note 6**).

3. Start the reaction by adding 3 units of phosphoglyceric phosphokinase (PPK), and then monitor the reduction in absorbance after NADH oxidation at 340 nm for 10 min using a spectrophotometer linked to a chart recorder.

4. For monitoring the effect of $H_2O_2$ on this activity, add different concentrations of $H_2O_2$ to the reaction mixture for 5 min, followed by the addition of PPK. Monitor the activity for 10 min (*see* **Note 7**).

5. Repeat the experiment at least three times with either different batches of enzyme or plant extract.

6. To show that the effect of $H_2O_2$ is actually reversible by using a reducing agent, reversal of inhibition of GAPDH activity is performed by starting the reaction with PPK, adding $H_2O_2$ for 10 min, after which glutathione (10 m$M$) or dithiothreitol (10 m$M$) is added and activity continued to be monitored (*see* **Note** 8).

The results shown in **Fig. 7.1** were obtained from cell culture extracts of *Arabidopsis* by the use of these procedures. To confirm that the activity from the extract is indeed from the GAPDH protein identified via proteomics, molecular biology experiments need to be performed, which will involve the cloning and expression of the GAPDH protein, followed by purification of the pure protein. Subsequently, experiments such as those mentioned previously can be performed on this pure plant protein. In addition to identify the exact post-translational modification that may have occurred on the protein by $H_2O_2$, mass spectrometry can be used.

### 3.6. Growth and Analysis of Arabidopsis Plants

To study the effect of Cys-containing mutations (or any mutation) in plant proteins, one can make use of the different resources available for such purposes within the *Arabidopsis* plant community. A number of researchers have donated seed stocks of *Arabidopsis* to a national seed stock center, from where, for a small price, seeds can be obtained for noncommercial purposes by any researcher (*see* www.nasc.org). With a knowledge of what plant is needed, the holdings of the stock center can be searched for various seeds that have been donated. For example, the ethylene receptor ETR1 has a number of mutants that are available. Of this, the *etr1-1* mutant has a mutation in a particular cysteine residue, Cys65, which is required for ethylene binding. By hypothesizing that this Cys65 might be essential for ROS sensing and signalling,

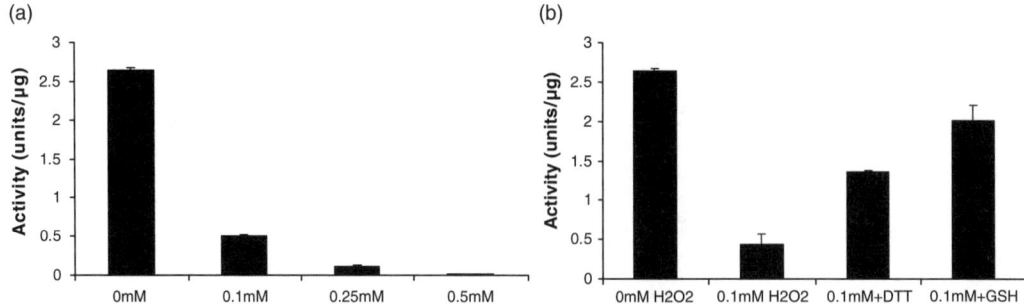

Fig. 7.1 Hydrogen peroxide ($H_2O_2$) inhibition of GAPDH activity is reversible. GAPDH was pretreated with $H_2O_2$ at different concentrations before the assay **(a)**. After GAPDH had been treated with $H_2O_2$ for 5 min, 10 m$M$ dithiothreitol (DTT) or reduced glutathione (GSH) was added and rates calculated after a further 10 min **(b)**.

various experiments were designed to test this. Seeds of *etr1-1* mutant plants were obtained from the stock center. Simultaneously, the corresponding wild type background plant seeds were also obtained.

***3.6.1. Growth of Arabidopsis Plants***

1. When the seeds have been obtained, store them at 4°C for at least 48 h, to break the dormancy of the seeds.
2. Sow both wild type and mutant *etr1-1* (mutant) seeds in separate trays of Levington's F2 compost with sand, after wetting the compost until it forms clumps in the hand (*see* **Note** 9). Sow the seeds evenly spaced out, and singly, using a toothpick with a wet end. On average approximately 15–18 plants can be sown on a 30 × 20-cm tray.
3. Place the trays in a controlled environment growth cabinet with the following settings: 22°C, 80% relative humidity, and 16-h photoperiod with lights of intensity 60–100 µE/m²/s (*see* **Note** 10).
4. Water the trays from below until the bottom of the tray is just above covered level (*see* **Note** 11). Water the plants three times a week, Monday, Wednesday, and Friday, at similar times, to maintain uniformity in plant growth.
5. Let the plants grow for at least 4 weeks until fully expanded rosette leaves have been obtained. Use leaves of similar size from both wild type and mutant plants.

***3.6.2. Phenotypic Analysis of Plants***

3.6.2.1. Stomatal bioassays

$H_2O_2$ causes stomatal closure in wild-type *Arabidopsis* leaves (4) and this can easily be used as a system to study the function of various signaling proteins. To perform stomatal bioassays, it is of absolute importance that both wild type and mutant plants are grown under exactly the same conditions, preferably in the same growth cabinet. This avoids any differences and variations due to growth conditions.

1. Detach single leaves from different plants that have been grown for 4 weeks (*see* **Note** 12).
2. Place each individual detached leaf in a 5-mL Petri dish with 2 mL of MES/KCl stomata bioassay buffer.
3. Cover the dishes and place them in the growth cabinet for 2.5 h.
4. Add various doses of $H_2O_2$ to different Petri dishes (wild type and mutant leaves).
5. Incubate for a further 2.5 h under the same conditions.
6. Blend each leaf individually in a Waring blender with 100 mL of water, for 30 s.
7. Collect the blended epidermal fragments on a 100-µm nylon mesh (SpectraMesh, BDH-Merck, UK).

8. Transfer the epidermal fragments to a glass microscope slide and place a cover slip on top.

9. Measure the stomatal apertures with the aid of an eyepiece graticule on a calibrated microscope. During measurements, make sure to move around and note apertures of at least 25 different stomata, from different fields of view. Also, ensure that the correct focal plane for the aperture is used. As the cell walls of guard cells are fairly thick, it can be difficult to focus on the inner walls of the aperture (*see* **Fig. 7.2**). Also, it is very important to repeat these experiments at least three to five times, preferably with different batches of plants.

Using this method the results in **Table 7.1** were obtained, which clearly indicate that with increasing concentrations of exogenous $H_2O_2$, stomata of wild-type guard cells respond by closing, whilst stomata of the mutant *etr1-1* plants do not close in response to $H_2O_2$. As the *etr1-1* mutant is mutated in the Cys65 residue this confirms that this residue is somehow involved in mediating $H_2O_2$-induced closure in *Arabidopsis* guard cells.

**3.6.2.2. Analysis of root growth**

Both exogenous and endogenous $H_2O_2$ are known to affect root growth responses in plants *(5)* and, therefore, one of the phenotypes that can be analysed in mutant versus wild-type plants is root growth in the presence of various doses of $H_2O_2$. To do this, experiments are performed on seedlings, so as to be able to monitor root growth easily. For this, seedlings are grown in tissue culture on nutrient agar plates.

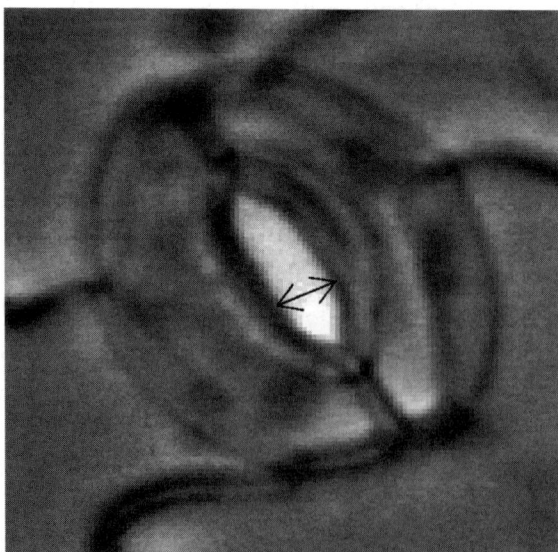

Fig. 7.2 Light microscope image of an open *Arabidopsis* stomate. Each cell surrounding the stomate is a guard cell. The arrow indicates the stomatal aperture (width) which is measured.

## Table 7.1
## Stomatal Aperture Measurements of Wild-Type (wt) and
## *etr1-1 Arabidopsis* l Response to H$_2$O$_2$ (*see* ref. *3*).

|  | Wt | SEM | etr1-1 | SEM |
|---|---|---|---|---|
| 0 | 2.64 | 0.094 | 2.98 | 0.097 |
| 10 | 2.05 | 0.094 | 2.81 | 0.094 |
| 50 | 1.63 | 0.088 | 2.76 | 0.077 |
| 100 | 1.83 | 0.078 | 2.57 | 0.088 |
| 200 | 1.39 | 0.213 | 2.56 | 0.084 |

Apertures are indicated in microns (µm); H$_2$O$_2$ concentrations in µm data are from five independent experiments. SEM = standard error of mean, $n$ = 10 guard cells.

1. First, prepare Murashige and Skoog medium full strength, with 0.5 g/L MES, pH 5.7, and with bacto-agar at a strength of 1.5% (w/v). Autoclave and cool until the medium bottle can be held in the hands.

2. In a laminar flow cabinet, pour the molten agar into Petri dishes (9 cm), either with or without H$_2$O$_2$ at various concentrations (*see* **Note** 13). When the agar has set, make sure the plates and lids are dry before sowing seeds onto them.

3. Surface sterilise a small amount of seeds (*see* **Note** 14) sufficient for 5–6 plates containing 15–20 seeds each. To surface sterilize, add 1 mL of 70% ethanol into each tube containing seeds and agitate gently at room temperature for 5 min.

4. Remove the ethanol by allowing the seeds to settle.

5. Add 1 mL of 10% bleach, 0.1% SDS solution and agitate gently for 5 min.

6. Remove the solution after 5 min and wash the seeds with sterile distilled water at least five times.

7. Resuspend the seeds in 1 mL of sterile water, ready to sow on the agar plates.

8. In a laminar flow cabinet, using a cut-off pipette tip and a 100-µL pipette, slowly pipet single seeds with water onto the surface of the agar plate, allowing approximately 1 cm between each seed. Sow the seeds in two horizontal rows, with a distance of approximately 3 cm between them.

9. Allow the plates to dry before sealing them with Parafilm.

10 Place the plates in the fridge for 2 d.

11. Subsequently incubate the plates for 1 week vertically on shelves in a controlled environment growth room (24°C, 16-h photoperiod with lights of intensity 100–150 µE/m²/s).

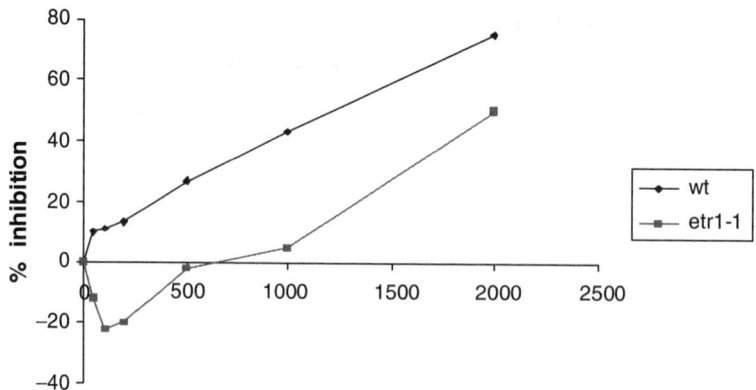

Fig. 7.3 Effect of $H_2O_2$ on root growth of wild-type Col-0 and *etr1-1* mutant seedlings. Seedlings were grown on MS agar plates in the presence or absence of increasing doses of $H_2O_2$, and root lengths measured after 7 days. The y-axis shows the percent inhibition of root growth compared to the controls for each plant. The x-axis shows increasing concentrations of $H_2O_2$ (in micromolar).

12. After 1 wk, observe root length by measuring the length of the roots from the base of the hypocotyl to the tip of the root.

With the use of this method, the root lengths of wild-type Col-0 and mutant *etr1-1 Arabidopsis* seedlings in the presence of $H_2O_2$ were measured and recorded, and examples of the results obtained are shown in **Fig. 7.3**.

## 4. Notes

1. Sterilize flasks by baking them at 200°C for 2 h.

2. The time of $H_2O_2$ treatment is determined empirically for different tissue types. The time mentioned here has been tested for cell cultures.

3. Volume of buffer to sample: 1.5-mL buffer: 3 g of fresh weight tissue, calculated by aspirating all media from cell cultures and weighing wet weight of cells in Falcon tube.

4. If for some reason the protein does not digest into peptides easily using trypsin, then alternative enzymes can be used (e.g., chymotrypsin).

5. For TCA/acetone precipitation, add ice cold 100% TCA to make a final concentration of 10% (v/v). Chill on ice for 15 min and centrifuge the sample for 20 min to remove the supernatant. Wash the pellet with 1 ml of 100% acetone, vortex well and re-centrifuge for 10 min. Repeat this process twice to wash the pellet. Air-dry the pellet and resuspend in the required buffer.

6. Soluble protein extract is obtained by homogenising the cells, followed by centrifugation at $12,000\times g$ for 10 min to obtain soluble proteins.

7. The activity of GAPDH can be followed by the oxidation of NADH as it is a stoichiometric reaction.

8. To confirm that the inhibition of GAPDH is by affects on the enzyme itself, rather than on the phosphoglyceric phosphokinase (PPK), the assay can be carried out in the reverse direction, using glyceraldehyde 3-phosphate as the substrate and following the reduction of NADH instead of its oxidation.

9. Over-wetting should be avoided.

10. It is very important to have the lights of the appropriate intensity as this vastly affects the growth of the plants.

11. Overwatering of plants must be avoided.

12. If the plants start flowering do not use them for the stomatal assays.

13. For getting accurate concentrations of $H_2O_2$ in each plate, make a $1,000\times$ stock of the required concentration of $H_2O_2$, and dilute this in 50 ml of medium in Falcon tubes, enough to pour two plates. Repeat with fresh tubes for each dose.

14. The amount of seeds in a microfuge tube which comes up to the 0.1 ml mark is approximately 300–500 seeds.

## References

1. Cooper, C., Patel, R. P., Brookes, P. S., and Darley-Usmar, V. M. (2002) Nanotransducers in cellular redox signalling: modification of thiols by reactive oxygen and reactive nitrogen species. *Trends Biochem Sci* **27**, 489–492.

2. Hancock, J. T., Henson, D., Nyirenda, M., Desikan, R., Harrison, J., Lewis, M., Hughes, J., and Neill, S. J. (2005) Proteomic identification of glyceraldehyde 3-phosphate dehydrogenase as an inhibitory target of hydrogen peroxide in *Arabidopsis*. *Plant Physiol Biochem.* **43**, 828–835.

3. Desikan, R., Hancock, J. T., Bright, J., Harrison, J., Weir, I., Hooley, R., and Neill, S. J. (2005) A novel role for ETR1: hydrogen peroxide signalling in stomatal guard cells. *Plant Physiol* **137**, 831–834.

4. Pei, Z.-M., Murata, Y., Benning, G., Thomine, S., Klusener, B., Allen, G. J., Grill, E., and Schroeder, J. I. (2000) Calcium channels activated by hydrogen peroxide mediate abscisic acid signalling in guard cells. *Nature* **406**, 731–734.

5. Foreman, J., Denidchik, V., Bothwell, J. H.F., Mylona, P., Miedema, H., Torres, M. A., Instead, P., Costa, S., Brownlee, C., Jones, J. D.G., Davies, J. M., and Dolan, L. (2003) Reactive oxygen species produced by NADPH oxidase regulate plant cell growth. *Nature* **422**, 442–446.

# Chapter 8

## Methods for Preparing Crystals of Reversibly Oxidized Proteins: Crystallization of Protein Tyrosine Phosphatase 1B as an Example

### Annette Salmeen and David Barford

### Abstract

Regulation of protein activity through the oxidation and reduction of cysteines is emerging as an important mechanism in the control of cell-signaling pathways. Protein tyrosine phosphatase 1B (PTP1B), for example, is reversibly inhibited by oxidation at the catalytic cysteine in response to stimulation of cells by insulin or epidermal growth factor. We have conducted structural studies on the redox regulation of PTP1B and have demonstrated that the oxidation of the catalytic cysteine results in the formation of a bond between the sulfur atom of the catalytic cysteine and the amide nitrogen of the neighboring serine. This bond, referred to here as a sulfenamide bond, is reversible upon the addition of glutathione, indicating that this sulfenamide intermediate could function within signaling pathways to protect the cysteine from overoxidation to less readily reducible states. Formation of the sulfenamide bond is accompanied by changes in the tertiary structure at the catalytic site, and these changes may be important for additional regulation of the enzyme. Here, we present methods for preparing crystals of PTP1B with a sulfenamide bond at the catalytic cysteine. The methods may be adaptable for other proteins that are subject to redox regulation.

**Keywords:** Protein tyrosine phosphatase, redox regulation, PTP1B, sulfenamide.

## 1. Introduction

Hydrogen peroxide ($H_2O_2$) has historically been considered to be toxic to cells and is associated with inducing DNA damage and cell stress. Mounting evidence, however, suggests that low levels of $H_2O_2$ can act as a second messenger in cell-signaling pathways by regulating the activity of enzymes, such as protein tyrosine phosphatases, protein kinases, and transcription factors (1–7).

John T. Hancock (ed.), *Methods in Molecular Biology, Redox-Mediated Signal Transduction, vol. 476*
© 2008 Humana Press, a part of Springer Science+Business Media, Totowa, NJ
DOI: 10.1007/978-1-59745-129-1_8

This "redox-regulation" primarily occurs via the reversible oxidation and reduction of cysteine residues, that is, through oxidation of cysteine residues to states that can be readily reduced, such as disulfide bonds (*see* **Table 8.1**). Further oxidation of cysteines to sulfonic acids ($Cys\text{-}SO_3$) can occur in vitro, but this is an unlikely state to occur in redox regulation of cell-signaling responses because sulfonic acids are not readily reduced. Protein crystallography studies on protein tyrosine phosphatase 1B (PTP1B) have revealed a novel reversibly oxidized state of the catalytic cysteine residue in which a bond is formed between the sulfur atom of the cysteine and the amide nitrogen of the neighboring serine residue (**Fig.** 8.1; refs. *10, 11*). Formation of this sulfenamide bond is accompanied by changes in the tertiary structure of PTP1B at the active site. These conformational changes may be significant for the regulation of the protein activity or localization because they expose a tyrosine residue that is susceptible to phosphorylation *(10)*.

In this chapter, we describe methods for preparing crystals of PTP1B with a sulfenamide bond at the catalytic cysteine and for establishing that the sulfenamide bond can be reduced with glutathione, a low-molecular weight thiol-containing peptide that is abundant in cells. We begin by assuming that a homogenous sample of purified protein has already been prepared and describe

## Table 8.1
## Oxidative Modifications of Cysteine Residues

| Name of cysteine modification | Chemical structure of modification | Reversibility | Time for formation in PTP1B crystals in excess $H_2O_2$ *(10)* |
|---|---|---|---|
| Thiolate anion | $Cys\text{-}S^-$ | N/A (already reduced) | N/A (already reduced) |
| Sulfenic acid | $Cys\text{-}SOH$ | DTT, GSH, other thiols or enzymes | Minutes–hours[a] |
| Sulfenamide bond | $Cys\text{-}S\text{-}N\text{-}Ser$[b] | DTT, GSH, other thiols or enzymes | Hours |
| Disulfide bond | $Cys\text{-}S\text{-}S\text{-}Cys$ | DTT, GSH, other thiols or enzymes | N/A (not formed in PTP1B) |
| Sulfinic acid | $Cys\text{-}SO_2$ | Reduction possible for some proteins with enzymatic activity *(8, 9)* | Hours–day |
| Sulfonic acid | $Cys\text{-}SO_3$ | No known examples of reversibility in biological systems | Days |

[a] Intermediate not observed in our studies, but estimated based on timing for sulfenamide bond formation
[b] Bond forms to a serine in PTP1B but may form to another amino acid in other proteins.

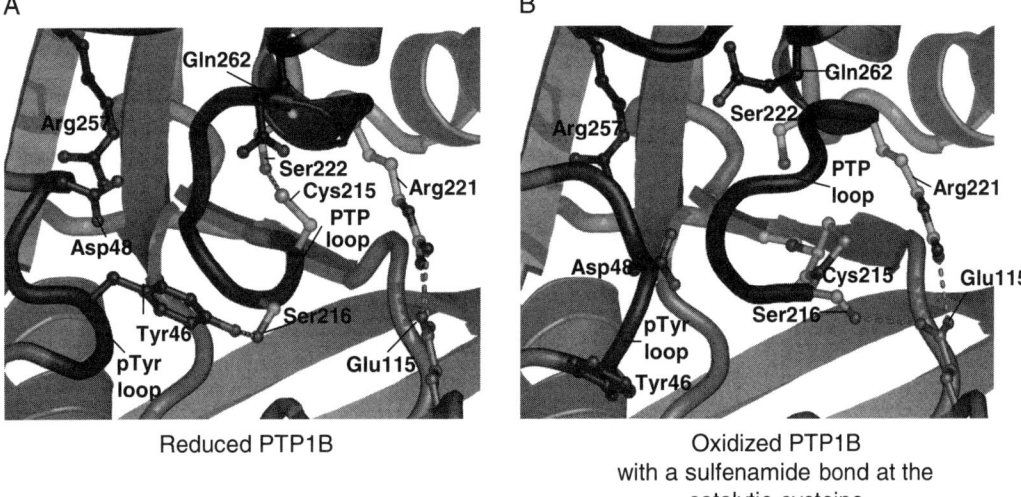

Fig. 8.1 Structures of the catalytic site of reduced **(A)** and oxidized **(B)** PTP1B are shown highlighting the structural changes that occur upon oxidation and formation of the sulfenamide bond. The conformations of two loops in the catalytic site of PTP1B, the PTP loop and pTyr recognition loop undergo dramatic conformational changes *(10)*.

two methods for preparing the crystals. In one method, the protein is oxidized in solution before crystallization. In the second method, crystals of reduced PTP1B are grown, and the protein is oxidized to the sulfenamide state while in the crystal. For PTP1B, both methods yielded similar crystal structures. However, if these methods are applied to another protein where the redox regulatory site occurs near a point of crystal contact, in-crystal oxidation may damage the crystal or inhibit a conformational change. In these cases, in-crystal oxidation may not be possible.

## 2. Materials

This method requires that a minimum of 0.5 mg of homogenous soluble purified protein has already been prepared (*see* **Note 1**). The purification procedures for the catalytic domain of PTP1B followed those previously described with the exception that Buffer B in **Table 1** contained 25 mM $NaH_2PO_4$ *(12)*.

### 2.1. Crystallization of PTP1B After Oxidation in Solution

#### 2.1.1. Oxidation of PTP1B in Solution

1. A minimum of 0.5 mg of purified PTP1B or 20 μg of purified protein for each crystallization condition that will be tested (*see* **Note 1**).

2. (Optional) *p*-NPP assay buffer: 10 m*M* Tris-HCl, pH 7.5; 25 m*M* sodium chloride; 0.2 m*M* ethylenediamine tetraacetic acid (EDTA; *see* **Note 2**).

3. (Optional) 200 mM *para*-nitrophenyl phosphate (*p*-NPP). The solution can be stored at –20°C.

4. Hydrogen peroxide (purchased as a 30% solution; *see* **Note 3**).

5. Dialysis tubing and clips.

6. Oxidized PTP1B crystallization buffer: 4 L of 10 m*M* Tris-HCl, pH 7.5; 25 m*M* sodium chloride; and 0.2 m*M* ethylenediamine tetraacetic acid (EDTA) (*see* **Note 2**).

7. 0.2-µ*M* filters for degassing buffers.

8. Bradford assay reagent for determining protein concentrations.

*2.1.2. Crystallization of PTP1B That Has Been Oxidized in Solution*

1. Purified concentrated oxidized PTP1B prepared as described in Section 3.1.1.

2. Sterile 50-mL containers for storing crystallization buffers (Falcon tubes, for example).

3. Crystallization stock solutions (prepare 50 mL of each solution and store at room temperature):
   a. 1 *M* *N*-(2-hydroxyethyl)piperazine-*N'*-2-ethanesulfonic acid (HEPES), pH 7.0.
   b. 1 *M* HEPES, pH 7.5.
   c. 1 *M* HEPES, pH 8.0.
   d. 40% (w/v) polyethylene glycol 8000 (PEG8K).
   e. 2 *M* magnesium acetate (MgOAc).
   f. 2 *M* EDTA, pH 8.0 (*see* **Note 2**).

4. 0.2-µm filter for syringes.

5. 10-, 25-, or 50-mL syringes.

6. Pre-greased Linbro 24-well plate with cover.

7. 22-mm silanized glass coverslips (enough for every crystallization condition that will be tested and extras in case they break; Hampton Research or Molecular Dimensions Limited).

## 2.2. In-Crystal Oxidation and Reduction of PTP1B

### 2.2.1. Crystallization of Reduced PTP1B

1. A minimum of 0.5 mg of PTP1B that has been purified in the presence of 5 m*M* dithiothreitol (DTT).

2. Reduced PTP1B crystallization buffer: 10 m*M* Tris-HCl, pH 7.5; 25 m*M* sodium chloride; 0.2 m*M* ethylenediamine tetraacetic acid (EDTA).

3. 0.5 *M* DTT prepared freshly prepared (*see* **Note 4**).

4. **Items 2–7** from Section 2.1.2.

### 2.2.2. In-Crystal Oxidation of PTP1B

1. At least two crystals of reduced PTP1B prepared as described in Section 3.2.1. Starting with four crystals is recommended (two crystals for oxidation and two crystals for reduction; *see* **Note 5**).

2. Empty pre-greased Linbro 24-well plate.

3. At least two micro-bridges for sitting drop crystallization (available from Hampton research, these can be cleaned and reused).

4. Vacuum grease.

5. Silanized glass coverslips that can be used to seal the wells of the Linbro-plate.

6. Crystal wash buffer: 0.1 $M$ Hepes, pH 7.5; 0.2 $M$ MgOAc; 0.2 m$M$ EDTA; and PEG8K ranging from 12% to 17% (*see* **Note 6**).

7. Crystal oxidation buffer: 0.1 $M$ HEPES, pH 7.5; 0.2 $M$ MgOAc; 0.2 m$M$ EDTA; and PEG8K ranging from 12% to 17%, 20 μ$M$ H$_2$O$_2$ (*see* **Note 6**).

8. Tweezers.

9. Mounted cryoloop used for transferring crystals from the hanging drop into the wash buffer.

10. (Optional) *p*-NPP assay buffer: 10 m$M$ Tris-HCl pH 7.5, 25 m$M$ sodium chloride, 0.2 m$M$ (*see* **Note 2**).

11. (Optional) 200 mM *p*-NPP (can be stored at –20°C).

*2.2.3. In-Crystal Reduction of PTP1B*

1. All items from Section 2.2.2 except the reduced crystals (**item 1** in Sect. 2.2.2) and the crystal oxidation buffer (**item 7** in Section 2.2.2).

2. A minimum of one crystal of oxidized PTP1B prepared as described in Section 3.1.2. Having four crystals of oxidized PTP1B is recommended (*see* **Note 5**).

3. Crystal reduction buffer: 0.1 $M$ HEPES, pH 7.5; 0.2 $M$ MgOAc; 0.2 m$M$ EDTA; PEG8K ranging from 12% to 17%(w/v); and either DTT (5 m$M$) or GSH (5 m$M$; *see* **Notes 4, 6**, and **7**).

***2.3. Freezing Crystals to Trap Them in an Oxidation State at a Certain Time and to Prepare Them for Structure Determination***

1. Cryoprotectant buffer: 0.1 $M$ HEPES, pH 7.5; 0.2 $M$ MgOAc; 0.2 m$M$ EDTA; 17.5% (v/v) 2-methyl-2,4-pentanediol; and 18–20% (w/v) PEG8K (*see* **Note 8**) *(13)*.

2. Microbridge for sitting-drop crystallization.

3. Tweezers.

4. Mounted cryoloops (Hampton research).

5. Dewar filled with liquid nitrogen (~25 L).

6. X-ray diffraction facility with cryostream and goniometer head mounted with an arc (optional, *see* **Note 9**).

7. Cryovials.

8. Cryocane that will hold the cryovials.

9. Crystal wand, Crystal Tongs or other device to transfer mounted cryoloops into the cryovials.

# 3. Methods

## 3.1. Crystallization of PTP1B After Oxidation in Solution

### 3.1.1. Oxidation of PTP1B in Solution

#### 3.1.1.1. (Optional) Protein Tyrosine Phosphatase Activity Assay

Protein tyrosine phosphatases will hydrolyze *para*-nitrophenyl phosphate (*p*-NPP) to *para*-nitrophenol (*p*-NP) when they are reduced and the rate of production of *p*-NP can be monitored on a spectrophotometer by measuring the absorbance at 405 nm *(14)*. This assay provides a rapid means to test the hydrogen peroxide solution, the protein sample, and the procedures on a small aliquot of protein before spending time and reagents to prepare the protein for crystallization.

1. Prepare a mixture of 1 µ$M$ PTP1B and *p*-NPP assay buffer to 990 µL in a 1.5-mL micro-centrifuge tube.

2. Prepare a second mixture of 1 µ$M$ PTP1B, 50 µ$M$ $H_2O_2$ and *p*-NPP assay buffer to 990 µL in a 1.5-mL micro-centrifuge tube (*see* **Note 10**).

3. Incubate the assay mixtures at room temperature for 20 min (*see* **Note 10**).

4. While waiting, prepare a blank for the assay by adding 10 µl of 200 mM *p*-NPP to 990 µL of *p*-NPP assay buffer.

5. Place the blank in a spectrophotometer and monitor the absorbance at 405 nm for 5 min.

6. After the samples have been incubated for 20 min add 10 µL of 200 mM *p*-NPP to one sample. If a spectrophotometer is available that can monitor the absorbance of multiple samples simultaneously then *p*-NPP can be added to both samples and **steps 7–9** can be done at the same time (*see* **Note 11**).

7. Transfer the sample to a plastic cuvette.

8. Immediately after the addition of *p*-NPP to the first sample place the sample in a spectrophotometer and monitor the absorbance at 405 nm for 5 min (*see* **Note 12**).

9. The degree of oxidation can be assessed by looking at a graph of the absorbance at 405 nm vs time. For the sample in which PTP1B was incubated in 50 µ$M$ $H_2O_2$ for 20 min, there should be very little or no change in absorbance at 405 nm. For the sample without $H_2O_2$, there should be an observable change in absorbance at 405 nm. *See* **Note 13** for troubleshooting ideas if either the oxidized or reduced sample were not as expected.

#### 3.1.1.2. Preparation of Oxidized PTP1B

1. Filter, de-gas, and chill 4 L of oxidized PTP1B crystallization buffer to 4°C (*see* **Note 14**).

2. Transfer the PTP1B sample to either dialysis tubing or a dialysis slide with a molecular weight cutoff less than the molecular weight of the protein (37 kDa).

3. Place the tubing or dialysis bag into 2 L of chilled oxidized PTP1B crystallization buffer. Add a magnetic stirring rod to the buffer and incubate the sample at 4°C for 2 h while stirring.

4. After 2 h, transfer the dialysis tubing or slide into 2 L of fresh chilled oxidized PTP1B crystallization buffer and place at 4°C for 2 additional hours while stirring.

5. To determine the protein concentration, measure the absorbance of the protein sample at 280 nm and calculate the concentration of the protein using Beer's law ($A = \varepsilon l c$ where $\varepsilon$ is equal to the extinction coefficient of the protein (66,610 M/cm), l is the length of the cuvette (usually 1 cm) and $c$ is the concentration).

6. Add $H_2O_2$ such that the final ratio of $H_2O_2$ to protein is 1:1.25 (*see* **Note 15**).

7. Incubate the mixture for 50 min at room temperature to allow for the oxidation to take place.

8. Concentrate the protein using an Amicon Centricon concentrator with a 10,000 molecular weight cut off to approximately 8.5 mg/mL (*see* **Note 16**). The final volume needed to obtain a concentration of 8.5 mg/mL can be estimated based on the concentration measurement in **step 5**.

9. Check the final protein concentration by either measuring the absorbance of the protein at 280 nm or by doing a Bradford assay to determine protein concentration (*see* **Note 17**).

*3.1.2. Crystallization of PTP1B That Has Been Oxidized in Solution*

PTP1B is crystallized here using a standard hanging drop method for protein crystallization *(15)*. Most of the preparation for crystallizing the protein is done in a cold room since this protein crystallizes at 4°C.

1. Calculate the volumes of the stock solutions that will be needed in each well (typical total volume per well is 1 mL) for the crystallization conditions that will be tested. For PTP1B the crystallization conditions that were tested were 0.1 *M* HEPES, pH 7.0, 7.5, and 8.0; 0.2 *M* MgOAc; 0.2 m*M* EDTA; and PEG ranging from 12% to 17% *(12)*.

2. Prepare 50 mL of crystallization stock solutions (*see* Sect. 2.1.2a–f).

3. Filter the stock solutions through a 0.2-μm syringe filter into a sterile plastic tube (for example, FALCON tubes available from VWR or Fischer).

4. Dispense volumes of each reagent required for crystallization into the wells of a 24-well pre-greased Linbro plate and mix well.

5. Chill the crystallization plate at 4°C. Also chill the silanized glass cover slips (enough for 1 coverslip/well plus several extras in case they break; *see* **Note 18**).

6. After the solutions and coverslips have reached 4°C the crystallization plates can be set up. For the first crystallization well mix together 2 µL of the well solution (from **steps 4** and **5** of Sect. 3.1.2) with 2 µL of the protein solution (prepared in Sect. 3.1.1.2) in the middle of a silanized glass coverslip.

7. Mix the 4-µL drop on the coverslip by gentle pipetting several times.

8. Invert the glass coverslip and place on the first crystallization well aligned with the well opening.

9. Repeat **steps 7** and **8** for all of the wells in the crystallization plate.

10. Ensure that each coverslip is well sealed over the crystallization well. Pressure should be applied uniformly over the coverslip such that any gap that was left in the grease is closed.

11. Leave the crystallization plate at 4°C and periodically check for crystal formation using a dissecting microscope with approximately 40× magnification. Typical crystal formation can take anywhere from 2 weeks to 1 month and for this particular protocol it is recommended to freeze the crystals shortly after they have formed (*see* **Note 19**).

### 3.2. In-Crystal Oxidation and Reduction of PTP1B

#### 3.2.1. Crystallization of Reduced PTP1B

1. Prepare reduced PTP1B for crystallization following **steps 1–5** listed in Section 3.1.1.2, but use the reduced PTP1B crystallization buffer instead of the oxidized PTP1B crystallization buffer.

2. Concentrate PTP1B using an Amicon centricon concentrator with a 10,000 molecular weight cut off to 10 mg/mL (*see* **Note 16**). Estimate the final volume based on the initial concentration measurement, and check the final concentration using a Bradford protein assay.

3. Follow the crystallization procedures described in Section 3.1.2 except add 5 m$M$ DTT (10 µL /well of a 0.5 $M$ stock solution) to each of the crystallization wells.

#### 3.2.2. In-Crystal Oxidation of PTP1B to a Sulfenamide Bond at the Catalytic Cysteine

1. Place micro-bridges for sitting-drop crystallization into two wells of an empty Linbro 24-well plate.

2. Add 40 µL of crystal wash buffer into one of the micro-bridges.

3. Add 4 µL of crystal oxidation buffer into the empty micro-bridge (*see* **Note 20**).

4. Remove the silanized glass coverslip from the well with the reduced PTP1B crystals that are going to be oxidized (*see*

**Note 21**), and place it face up on the stage of a dissecting microscope with approximately 50× magnification.

5. Grasp the base of a mounted cryoloop with a pair of forceps and use the cryoloop itself to remove a single crystal from the drop that the crystal of interest has been growing in. This can be difficult at first, but becomes easier with practice.

6. Plunge the tip of the cryoloop (the circular region) into the crystal wash buffer that has been prepared in the microbridge. Swipe the loop in the buffer several times to ensure that the crystal is released from the loop. By placing the microbridge (either in the Linbro 24-well dish or without whichever is more convenient) on the microscope stage, this step can be observed using the microscope to ensure that the transfer is successful. This step washes the DTT reducing agent out of the crystal which would inhibit the oxidation of the crystal in the subsequent step.

7. Using the cryoloop, transfer the crystal from the micro-bridge with the crystal wash buffer to the microbridge with the 4 µL of crystal oxidation buffer. Again, check the well under the microscope to ensure that the crystal transfer was successful.

8. Repeat **steps 5–7** with three additional crystals (*see* **Note 22**).

9. Using a clean glass coverslip, seal the well of the Linbro-24 well plate in order to prevent evaporation of the drop during the oxidation.

10. (Optional) The crystals of PTP1B can be checked to see if the protein has been completely oxidized. Redissolved crystals of PTP1B will hydrolyze *p*-NPP in the phosphatase activity assay described in Section 3.1.1.1. In **step 1**, the crystal is redissolved in 990 µL of buffer. After that the assay proceeds as described in Section 3.1.1.1. If the absorbance at 405 nm does not change then the protein is oxidized.

11. After 12–16 h, remove at least one crystal to freeze in its oxidized state following the procedures in Section 3.3. For the other three crystals proceed to Section 3.2.3.

*3.2.3. In-Crystal Reduction of Oxidized PTP1B*

The crystals can be reduced using solutions of either DTT or GSH. Either one can be added to the crystallization buffer solution as the reducing agent.

1. Place micro-bridges for sitting-drop crystallization into two wells of an empty Linbro 24-well plate.

2. Add 40 µL of crystal wash buffer into one of the micro-bridges.

3. Add 40 µL of crystal reduction buffer to the empty micro-bridge.

4. Remove the silianized glass coverslip from the well with the oxidized PTP1B crystals that are going to be reduced and place either the microbridge containing the oxidized crystals or the entire Linbro 24-well dish on the microscope stage.

5. Grasp the base of a mounted cryoloop with a pair of forceps and use the cryoloop itself to remove a single crystal of oxidized PTP1B from the drop in the microbridge.

6. Transfer the crystal into the crystal wash buffer as described in Sect. 3.2.2.6. This washes any remaining $H_2O_2$ out of the crystal, which would inhibit the reduction of the crystallized protein.

7. Using the cryoloop, transfer the crystal from the microbridge with the crystal wash buffer to the microbridge with the 40 µL of crystal reduction buffer. Again, check the well under the microscope to ensure that the crystal transfer was successful.

8. Repeat **steps 5–7** with any additional crystals.

9. Using a clean glass coverslip, seal the well of the Linbro-24 well plate containing the microbridge with the crystal reduction buffer and the freshly transferred crystals in order to prevent evaporation of the drop during the reduction.

10. (Optional) After a period of time, one of the crystals that has been soaked in the crystal reduction buffer can be checked to see whether the protein activity has been restored. Redissolved crystals of PTP1B will hydrolyze *p*-NPP in the phosphatase activity assay described in Section 3.1.1.1. In **step 1**, the crystal is redissolved in 990 µL of buffer. After that the assay proceeds as described in Section 3.1.1.1. If the absorbance at 405 nm changes, the activity of the protein has been at least partially restored.

11. After 48 h remove at least one crystal from the microbridge with the crystal reduction buffer to determine the structure of the reduced protein and follow the procedures in Section 3.3 for freezing the crystal.

### *3.3. Freezing Crystals to Trap Them in an Oxidation State and Prepare Them for Structure Determination*

Conducting crystallographic experiments at cryogenic temperatures of approx 100 K can reduce the radiation damage that occurs to protein crystals when they are exposed to x-rays *(16–18)*. In addition, flash-cooling of crystals for the experiments described here was an essential step for trapping PTP1B in its sulfenamide state and for collecting time series data for PTP1B oxidation. Freezing of the crystals dramatically slows the rate of reaction of the crystallized protein with atmospheric oxygen and essentially traps the crystallized protein in the oxidation state it was in at the time it was frozen.

Crystals can be frozen either by placing them directly into a cryostream at an x-ray diffraction facility or by immersing them into liquid nitrogen. The advantage of a cryostream is that one can check the cryoprotectant buffer (*see* Sect. 3.3.4), ensure that the crystals are successfully transferred into the cryoloops, and check the diffraction limit of the crystals prior to storing them. This protocol describes the procedures for freezing crystals in a cryostream but the protocol can be adapted to freezing crystals by immersing them in liquid nitrogen.

1. Ensure that a sufficient amount of liquid nitrogen is available for freezing the crystals and for storing the crystals if data will not be collected immediately. If the crystals are going to be stored, carefully fill a ~25 L dewar with liquid nitrogen. Liquid nitrogen should be handled with care following appropriate safety procedures.

2. Prepare 10 mL of cryoprotectant buffer (*see* **Note 23**).

3. Put 40 µl of cryoprotectant buffer into a clean empty sitting-drop microbridge.

4. (Optional) Using a pair of tweezers, grasp the base of a mounted cryoloop and dip the cryoloop into the cryoprotectant buffer. Mount the cryoloop into a cryostream (*see* **Note 24**) and ensure that the buffer in the loop looks like a clear glassy solution and that ice crystals have not formed. If ice has formed, the cryoprotectant buffer must be adjusted starting by adding a 1–2% higher concentration of PEG (*see* **Note 23**).

5. Position both the microbridge with the cryoprotectant buffer and the microbridge with the crystals to be frozen, on the stage of a dissecting microscope. Bring the crystals into focus in the microscope and grasp the base of a mounted cryoloop with a pair of tweezers.

6. While looking at the crystals through the microscope, use the cryoloop to remove a single crystal from the drop where the crystals have been growing and transfer the crystal to the microbridge containing the cryoprotectant buffer. Gently swipe the cryoloop back and forth in the cryoprotectant buffer until the crystal is transferred from the cryoloop and fully immersed in the cryoprotectant buffer.

7. Use the cryoloop, to pick up the crystal again and remove the crystal from the cryoprotectant buffer (*see* **Note 25**). Mount the crystal in the cryostream (*see* **Note 24**).

8. (Optional) Once the crystal is mounted in the cryostream one can check the video-camera to ensure that the crystal has been successfully transferred into the cryoloop. At this step it can also be informative to collect an initial diffraction image to determine the diffraction limit of the crystal. The crystal then

can either be transferred to a dewar of liquid nitrogen for storage or used immediately for data collection. Data collection techniques will not be discussed in this chapter so the reader is referred to other references that describe methods for crystallographic data collection and structure determination.

9. If the crystal is going to be stored, first place an empty cryovial in the cryovial holder and immerse it in liquid nitrogen until bubbles are no longer being released from the cryovial.

10. To transfer the crystal, use the cryovial-holder to raise the cryovial out of the liquid nitrogen and position the cryovial such that the cryoloop is suspended in the liquid nitrogen and the base of the cryoloop is essentially positioned as a cap over the cryovial. Slide the cryoloop off where it is mounted. The cryoloop should then be suspended in the liquid nitrogen in the cryovial.

11. Using a pair of cryogloves transfer the cryovial to the cryocane and place the cryocane in a storage dewar that is filled with liquid nitrogen. The crystals can be stored for at least several months provided that the dewar is kept filled with liquid nitrogen.

## 4. Notes

1. One 24-well plate requires ~0.5 mg of PTP1B, which crystallizes at a concentration of 10 mg/mL. More protein may be needed depending on the number of crystallization conditions that will be tested.

2. EDTA is insoluble in solution until the pH reaches approximately 8.

3. Hydrogen peroxide is light sensitive and will degrade over time. It should be stored in the dark and purchased in small quantities. For these procedures it is recommended that the $H_2O_2$ is not more than 6 months old. The reagent can be tested following the procedures in Section 3.1.1.1.

4. DTT is a reducing agent. It will react with cysteine residues within a protein and keep the cysteine residues in a thiolate anion state (*see* **Table 8.1**). It is more susceptible to oxidation once it is dissolved in aqueous solutions and should therefore either be prepared freshly on the day of use or stored in aliquots at –20°C.

5. Even within a single crystallization well there is variability in the quality of the crystals. Preparing several crystals at a

time improves the probability of finding a crystal that will diffract well in the crystallography experiment. In addition, the crystals will be transferred to several different buffers and can be damaged or lost as they are transferred. Preparing several crystals at once requires minimal extra time and again improves the probability of having a high quality crystal for the x-ray diffraction experiment.

6. The concentration of PEG in the crystal wash buffer, the crystal oxidation buffer or the crystal reduction buffer should match or slightly exceed the concentration of PEG that was present in the well solution while the crystal was growing. For example, crystals that grew in 12% PEG 8K should have at least 12% PEG 8K in the various buffers.

7. Glutathione can become oxidized rapidly once it is dissolved in solution. This oxidation also occurs slowly in the solid state. Liquid solutions should be made up freshly on the day of use. It is recommended that the bottles of glutathione not be more than 1 yr old.

8. The appropriate cryoprotectant buffer is dependent upon the conditions that were used to grow the crystal. For PTP1B crystals that grow in relatively high PEG concentrations (16–17%) a cryoprotectant buffer with 20% PEG is more suitable. For those that grow in lower PEG concentrations (12–15%) a cryoprotectant buffer with 18% PEG is more suitable.

9. While freezing crystals, it is convenient to be able to check that the crystal is in the cryoloop and to check how well it diffracts. However, if no cryostream is available, the crystals can be frozen and stored directly in the liquid nitrogen without the need for a cryostream or goniometer head. If the crystals are being frozen immediately prior to data collection they can be frozen in the cryostream and data collection can be initiated immediately.

10. This assay can also be used to test different concentrations of hydrogen peroxide or different times of oxidation.

11. The concentration of $p$-NPP can be adjusted such that the absorbance at 405 n$M$ is detectable and linear for most of the 5-min assay.

12. If the spectrophotometer is not set up to continuously record the absorbance versus time then the absorbance can be recorded at 30-s intervals.

13. If a change in absorbance at 405 nm is observed for the sample with $H_2O_2$ (i.e., the protein is still active even though $H_2O_2$ was added), then it may be necessary to use a longer oxidation time or to use a fresh supply of $H_2O_2$. If no change in absorbance at 405 nm is observed for the sample without $H_2O_2$, then

it may mean that the protein sample is not properly reduced. The assay could be repeated after treating the sample with freshly made 5 m$M$ DTT. If there is still no observable change in absorbance for the sample without $H_2O_2$, then a fresh protein sample may need to be prepared as the catalytic cysteine may have become oxidized to a sulfonic acid.

14. Proteins are usually purified with reducing agents present, but the reducing agents need to be removed before the addition of hydrogen peroxide. This dialysis step removes the reducing agents. A buffer exchange column could also be used for this step.

15. The selection of a ratio of 1:1.25 for the ratio of PTP1B to $H_2O_2$ is to ensure that there is enough $H_2O_2$ so complete oxidation can occur, but to prevent the further oxidation of the PTP1B catalytic cysteine to a sulfinic or sulfonic acid, or of non-catalytic cysteine and methionine residues.

16. Oxidized PTP1B forms precipitates in solution more rapidly than reduced PTP1B. At 8.5 mg/ml there was a small amount of protein precipitation, but the majority of the protein was still soluble. This precipitate was filtered and the protein was immediately placed into crystallization trials. Reduced PTP1B can be crystallized at a higher concentration (10 mg/mL).

17. Bradford protein assays require a smaller volume than measurements of the protein concentration at 280 nm and are convenient for checking protein concentrations if only a small amount of protein is available.

18. Some proteins such as PTP1B crystallize at 4°C whereas others crystallize better at greater temperatures. For proteins that crystallize at room temperature this step can be omitted.

19. The timing of crystallization and data collection is important. Sufficient oxygen from the air can enter crystallization wells and cause the reversible sulfenamide bond to become oxidized to a sulfonic acid. It is therefore important to freeze the crystals and collect data in a timely fashion to trap the readily reducible oxidized intermediate.

20. To trap PTP1B in the sulfenamide bond state, the crystals were soaked in a small volume of 4 µL with 20 µ$M$ of $H_2O_2$. This was done to limit the availability of $H_2O_2$ to prevent oxidation of the catalytic cysteine of PTP1B to sulfonic acid. The selection of 20 µ$M$ was determined based on estimating the total amount of PTP1B in the crystallization drop. After diluting the protein 1:2 in crystallization well buffer the final PTP1B concentration was 5 mg/ml. The molecular weight of the PTP1B catalytic domain is (37 kDa) which made the concentration of PTP1B in the crystallization drop approximately 130 µ$M$.

21. Glass coverslips can crack easily as they are removed. This can be minimized by gently rotating the coverslip to loosen the grease seal before prying it off. Also, although the external appearance of a protein crystal does not always correlate to how well it will diffract, larger regular shaped crystals tend to give a better diffraction pattern and were chosen at this step over smaller less regularly shaped crystals.

22. Four crystals were found to be suitable for this procedure. Preparing more crystals at this step may enhance chances for finding a crystal that diffracts well. However, adding too many crystals at this step may leave insufficient hydrogen peroxide for complete oxidation to the sulfenamide state. A third microbridge with an additional 4-μL drop of crystal oxidation buffer could be used to oxidize four additional crystals.

23. Cryoprotectant buffers help to minimize ice formation and crystal damage during the freezing process. For more information on selection and optimization of cryoprotectant buffers, *see* **refs.** *(16–18)*.

24. Mounting and dismounting crystals is a user- and facility-dependent procedure and detailed protocols are not presented here. Common procedures are the cryotong method (Hampton Research) and the detachable extended arc method (Molecular Dimensions Limited).

25. Handling of the crystals should be minimized as much as possible. To prevent the presence of excess cryoprotectant buffer surrounding the crystal, it is advantageous to pull the cryoloop from the cryoprotectant buffer such that the cryoloop is perpendicular to the meniscus of the solution. Also, the crystal should ideally be placed in the center of the cryoloop.

## Acknowledgments

We thank Nicholaas K. Tonks, Janik N. Anderson, Mike P. Meyers, John A. Hinks, and Harren Jhoti for helpful discussions. The work presented here was funded by grants from Cancer Research-UK (D.B.).

## References

1. Mahadev, K., Zilbering, A., Zhu, L., and Goldstein, B. J. 2001 Insulin-stimulated hydrogen peroxide reversibly inhibits protein-tyrosine phosphatase 1b in vivo and enhances the early insulin action cascade. *J Biol Chem* **276**, 21938–21942.

2. Lee, S. R., Kwon, K. S., Kim, S. R., and Rhee, S. G. 1998 Reversible inactivation of

protein-tyrosine phosphatase 1B in A431 cells stimulated with epidermal growth factor. *J Biol Chem* **273**, 15366–15372.

3. Finkel, T. 2003 Oxidant signals and oxidative stress. *Curr Opin Cell Biol* **15**, 247–254.

4. Finkel, T. 2000 Redox-dependent signal transduction. *FEBS Lett* **476**, 52–54.

5. Rhee, S. G., Bae, Y. S., Lee, S.-R., and Jaeyul, K. 2000 Hydrogen peroxide: key messenger that modulates protein phosphorylation through cysteine oxidation. *Sci STKE* **2000**(53), PE1.

6. Liu, H., Colavitti, R., Rovira, I. I., and Finkel, T. 2005 Redox-dependent transcriptional regulation. *Circ Res* **97**, 967–974.

7. Salmeen, A., and Barford, D. 2005 Functions and mechanisms of redox regulation of cysteine based phosphatases. *Antioxid Redox Signal* **7**, 560–577.

8. Woo, H.A., Chae, H. Z., Hwang, S. C., Yang, K. S., Kang, S. W., Kim, K., and Rhee, S. G. 2003 Reversing the inactivation of peroxiredoxins caused by cysteine sulfinic acid formation. *Science* **300**, 653–656.

9. Biteau, B., Labarre, J., and Toledano, M. B. 2003 ATP-dependent reduction of cysteine-sulphinic acid by *S. cerevisiae* sulphiredoxin. *Nature* **425**, 980–984.

10. Salmeen, A., Andersen, J. N., Myers, M. P., Meng, T. C., Hinks, J. A., Tonks, N. K., and Barford, D. 2003 Redox regulation of protein tyrosine phosphatase 1B involves a sulphenyl-amide intermediate. *Nature* **423**, 769–773.

11. van Montfort, R. L., Congreve, M., Tisi, D., Carr, R., and Jhoti, H. 2003 Oxidation state of the active-site cysteine in protein tyrosine phosphatase 1B. *Nature* **423**, 773–777.

12. Barford, D., Keller, J. C., Flint, A. J. and Tonks, N. K. 1994 Purification and crystallization of the catalytic domain of human protein tyrosine phosphatase 1B expressed in *Escherichia coli*. *J Mol Biol* **239**, 726–730.

13. Groves, M. R., Yao, Z. J., Roller, P. P., Burke, T. R., Jr. and Barford, D. 1998 Structural basis for inhibition of the protein tyrosine phosphatase 1B by phosphotyrosine peptide mimetics. *Biochemistry* **37**, 17773–17783.

14. Denu, J. M., Zhou, G., Wu, L., Zhao, R., Yuvaniyama, J., Saper, M. A., and Dixon, J. E. 1995 The purification and characterization of a human dual-specific protein tyrosine phosphatase [published erratum appears in *J Biol Chem* (1995) 270, 10358]. *J Biol Chem.* **270**, 3796–3803.

15. McPerson, A. 1999 *Crystallisation of Biological Macromolecules*. CSHL, New York.

16. Garmen, E. F. 1996 *Modern Methods for Rapid X-ray Diffraction Data Collection from Crystals of Macromolecules*. Methods in Mol Biol 56, 87–126.

17. Garman, E. F. and Schneider, T. R. 1997 Macromolecular crystallography. *J Appl Cryst* **30**, 211–237.

18. Hass, D. J. and Rossmann, M. G. 1970 Crystallographic studies on lactate dehydrogenase at –75°C. *Acta Cryst* **B26**, 998–1004.

# Color Plates

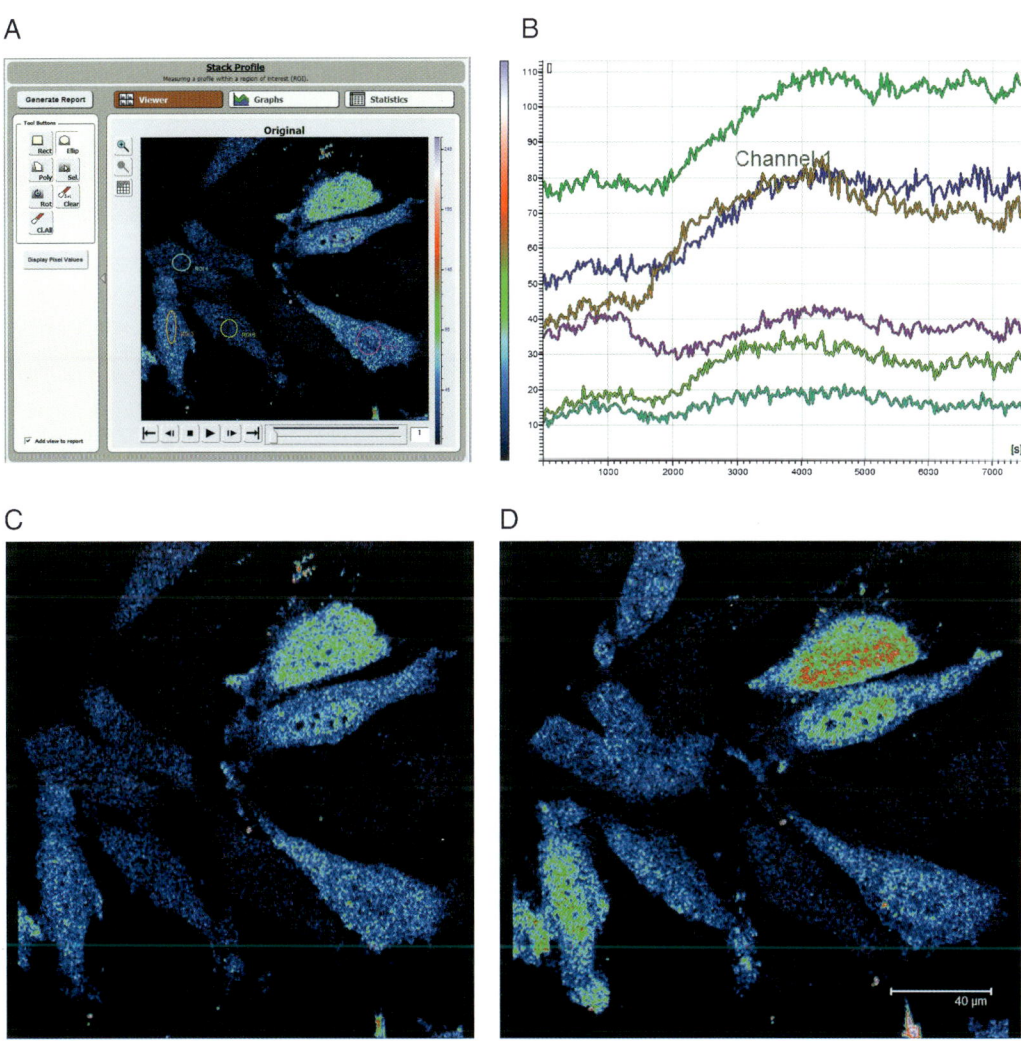

Fig. 6.1    Quantification of HyPer imaging series. **(A)** Screen shot from Leica Confocal Software "Stack Profile—Viewer" window. In this window all the stack can be viewed. The left upper buttons allow differently shaped regions of interest to be defined. By clicking the "Graphs" button above the image, time course **(B)** of signal intensities in selected regions of interest appears. Activating the right mouse button menu on the graphs allows the export of graphs to ".txt" or ". jpeg" formats. Pseudo-colored images of HeLa cells expressing HyPer-Cyto at the time point of **(C)** and 1 h after **(D)** EGF addition are shown.

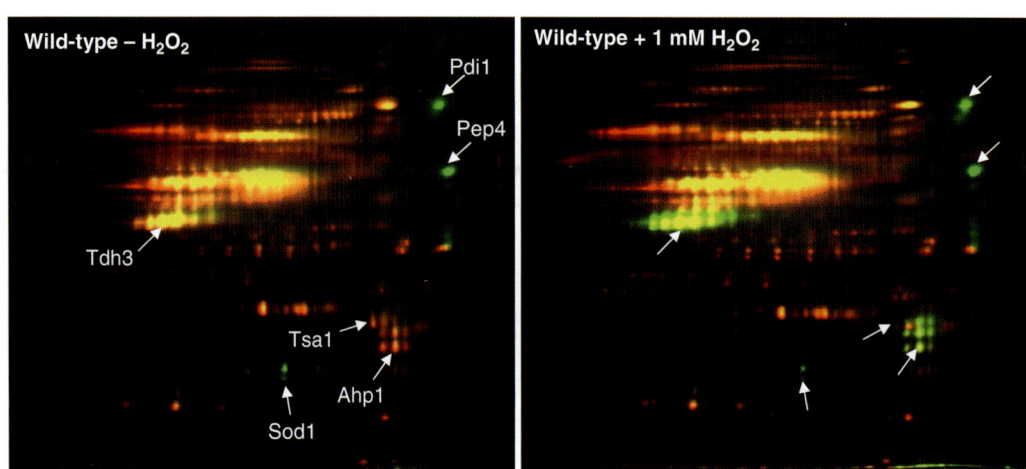

Fig. 13.4. 2D-gel electrophoresis of reduced and oxidized protein-thiols differentially labeled with fluorescent dyes of distinct emission wavelengths, as described in Sect. 3.2.2. Reduced thiols were first labeled with the NEM derivative DYE680 (*red*). Then, after their reduction by DTT, oxidized thiols were labeled with DYE780 (*green*). Red and green signal overlap appears in yellow. The gel on the left corresponds to untreated cells. Pdi1 and Sod1 are here fully oxidized and appear therefore in green. The gel on the right corresponds to cells treated with 1 m$M$ of $H_2O_2$ during 2 min. The green signals which appeared on the right gel correspond to protein-thiols oxidized by $H_2O_2$.

# Chapter 9

# Methods for the Study of Redox-Mediated Changes in p53 Structure and Function

## Kristine Mann

## Abstract

It is now generally accepted that both the structure and function of a number of specific transcriptional factors, including p53, are subject to redox regulation in cells in which these factors are expressed. The present chapter describes methods for the analysis of redox changes in the structure of p53 and the effect of redox modulation on binding of p53 to a DNA consensus sequence. In addition, methods are described for studying the effect of redox perturbations of cells on the functioning of p53 in the cell cycle and in apoptosis. By studying the effect of redox agents on p53, we have concluded that p53 is subject to structural redox modulation and that this modulation affects the functional ability of the protein to bind to DNA, to cause cell cycle arrest, and to trigger apoptosis.

**Keywords:** Redox regulation, p53 purification, structure, and function of p53, DNA binding, aminothiol WR1065 treatment, in vivo redox forms of p53 and thioredoxin, cell cycle analysis, apoptosis analysis.

---

## 1. Introduction

The redox state of specific cysteine residues in the p53 tumor suppressor protein plays a critical role in its ability to bind to consensus deoxyribonucleic acid (DNA) sequences and to function as a transcriptional regulator of target genes, including those involved in suppressing cellular transformation (1–3). Three cysteines (Cys-176, Cys-238, and Cys-242) along with His-179 are directly involved in binding of a zinc ion, which is essential for maintaining the active structural conformation of p53 (4). Other cysteines (Cys-124, Cys-135, Cys-141, Cys-275, and Cys-277) are located in the DNA-binding domain of p53 and may

John T. Hancock (ed.), *Methods in Molecular Biology, Redox-Mediated Signal Transduction, vol. 476*
© 2008 Humana Press, a part of Springer Science+Business Media, Totowa, NJ
DOI: 10.1007/978-1-59745-129-1_9

form a redox-sensitive site subject to modulation by the redox conditions within the cell *(1, 4)*. Oxidation–reduction of these cysteines would affect the conformation of the protein and in turn would affect the ability of p53 to arrest the cell cycle or trigger apoptosis in response to stress signals, either genotoxic or nongenotoxic in nature.

The conformation of p53 has been redox-modulated in vitro by treating purified p53 with the reducing agent dithiothreitol (DTT) or in vivo by treating p53-positive cells with a reducing agent such as aminothiol WR1065 or an oxidizing agent such as diamide. Redox changes in p53 structure have been analyzed by alterations in gel mobility. Structural redox perturbations of p53 have been shown to affect the ability of the protein to bind to consensus DNA in vitro and to function in G1 cell cycle arrest in vivo. In an alternative approach to studying the effect of redox on p53 function, p53-negative malignant cells were co-transfected with a plasmid expressing wild-type p53 and a plasmid +/− thioredoxin (TRX), and the effect of enhanced expression of the TRX on p53-induced apoptosis was examined. TRX is a small redox-active protein that helps maintain the overall redox state of the cell and control the redox status of important redox-sensitive transcription factors, such as NF-kB and AP1 *(5)*. The ability of p53 to induce apoptosis of the malignant cells was decreased in the presence of TRX, suggesting that the redox state of p53 plays a role in apoptosis.

## 2. Materials

### 2.1. Purification of Wild-Type p53 Protein

1. Sf9 (*Spodoptera frugiperda*) insect cells (ATCC, American Type Culture Collection)

2. TNM-FH medium (Grace's medium supplemented with yeastolate and lactalbumin hydrolysate; Sigma) + 10% fetal bovine serum (preheat serum at 56°C for 30 min before adding to the medium).

3. NPVp53, a recombinant baculovirus expressing wild-type murine p53 *(6)* (kindly provided by Peter Tegtmeyer).

4. Lysis buffer: 150 m$M$ Tris-HCl, pH 9.0; 150 m$M$ NaCl; 1 mM ethylene diamine tetraacetic acid (EDTA); 10% glycerol; 1% NP-40, 1 mg/mL leupeptin (Sigma); and 1 mg/mL aprotinin

5. Wash buffer C: 50 m$M$ Tris-HCl, pH 8.0; 500 m$M$ LiCl; 1 m$M$ EDTA—adjust pH to 8.0, then add glycerol to 10% final concentration, and autoclave

6. Wash buffer D: 20 m$M$ Tris-HCl, pH 8.5; 1 m$M$ EDTA—adjust pH, then add glycerol to 10% final concentration, and autoclave

7. Protein A-Sepharose beads were chemically crosslinked to the p53-specific monoclonal antibody PAb 421 by the method of Simanis and Lane *(7)* and stored in borate buffer.

8. p53 elution buffer: 50 m$M$ Tris-HCl, pH 8.0; 150 m$M$ NaCl; 1 m$M$ EDTA; 1 m$M$ DTT; and 0.1% NP-40 plus final concentration of 0.1 mg/mL of decapeptide KKGQST-SRHK-amide (Multiple Peptide Systems, San Diego, CA).

9. Chelex-100 resin (100–200 mesh, BioRad Laboratories, CA).

***2.2. Preparation of $^{32}P$ Labeled Target DNA***

1. pBS.KS$_{Shay}$ with the consensus sequence (5′-GGACAT-GCCCCGGGCATGTCC-3′) described by Funk et al. *(8)*, kindly provided by Dr. Judy Stenger.

2. Restriction endonucleases *Bam*HI and *Pst*I.

3. 1X TAE buffer: 40 m$M$ Tris-base, 1.1 mL of glacial acetic acid, 2 m$M$ EDTA (1X TAE buffer should have a final pH ~ 8.5).

4. DEAE membrane (Schleicher & Schuell NA45): wash for 10 min in 10 m$M$ EDTA, pH 7.6, followed by 5 min in 0.5 $M$ NaOH, and then wash several times with dH$_2$O before storing at 4°C.

5. High salt NET buffer: 20 m$M$ Tris-HCl, pH 8; 1 $M$ NaCl (increase up to 2.5 M if necessary); 0.1 m$M$ EDTA—autoclave solution.

6. Mixture (1:1:1) of 5 m$M$ dGTP, 5 m$M$ dCTP, and 5 m$M$ dTTP.

7. α-$^{32}$P-dATP (3000 Ci/mmol, 10 μCi/λ).

8. Klenow DNA polymerase.

9. Nuctrap (Stratagene) push column.

10. 1X STE buffer: 20 m$M$ Tris-HCl, pH 7.5; 100 m$M$ NaCl; 10 m$M$ EDTA.

11. Ethidium bromide (EtBr) stock solution: 5 mg EtBr/mL H$_2$O.

12. Low molecular weight Biomarker DNA (BioVentures, Inc.).

***2.3. Gel Mobility Shift Assay of p53 Bound to $^{32}P$-Labeled Target DNA***

1. Low-ionic-strength gel mix (100 mL): add 675 μL of 1 $M$ Tris-HCl, pH 7.9, 500 μL of 0.2 $M$ EDTA, pH 7.6; 330 μL of 1 $M$ sodium acetate, pH 7.8; 13.3 mL of 30% acrylamide; 2.5 mL of 2% bis-acrylamide; 5.0 mL of 50% glycerol; and 77.2 mL of dH$_2$O. Then filter through a 0.2-μ filter and store at 4°C until use.

2. $^{32}$P-end-labeled Shay DNA and Bluescript SK$^+$ DNA.

3. Purified wild-type p53 protein (at a concentration determined by Bradford assay), prepared with either chelexed (C) or non-chelexed (NC) solutions.

4. 2X DNA-binding buffer: 50 m$M$ N-(2-hydroxyethyl)piperazine-$N'$-2-ethanesulfonic acid (HEPES), pH 7.6; 100 m$M$ KCl; 1 mg/mL bovine serum albumin; 40% glycerol; and 0.2% NP-40.

5. 1X Tris-borate-EDTA buffer (TBE): 89 m$M$ Tris base, 89 m$M$ boric acid, 2 m$M$ EDTA.

6. Running buffer: 0.33X TBE + 0.1% Triton X-100, pre-chill at 4°C.

### 2.4. Blotting and Immunodetection of Mobility-Shifted p53-DNA Complexes

1. Nytran plus membrane (0.2 μ), maximum strength (Schleicher & Schuell).

2. TBS buffer: 15 m$M$ Tris-HCl, pH 7.8; 150 m$M$ NaCl; 1 m$M$ EDTA; 5 m$M$ NaN$_3$

3. Blocking buffer: TBS containing 0.1% gelatin and globulin-free BSA (Sigma) at a concentration of 100 μg/mL TBS.

4. TBGT: TBS + 0.1% gelatin + 0.1% Tween-20.

5. TBST: TBS + 0.1% Tween-20.

6. Substrate buffer: 100 m$M$ Tris-base, pH 9.5; 100 m$M$ NaCl, 50 m$M$ MgCl$_2$.

7. UV Stratalinker 2400 (Stratagene).

### 2.5. Effect of Redox Perturbing Agents, such as Aminothiol WR1065, Diamide, or Etoposide, on the Redox State of p53 and TRX in HCT116 Cells Treated with these Agents

1. Aminoguanidine (AG): make up a 10 m$M$ stock solution in culture medium (McCoy's medium + 10% fetal bovine serum) and pre-heat for 30 min at 56°C, with intermittent vortexing, to help get the AG into solution.

2. WR1065: make up as a 0.1 $M$ stock in PBS just before use or flush gently with argon if to be stored at 4°C before use, to prevent oxidation.

3. Diamide stock: 1.6 $M$ diamide in ddH$_2$O; store frozen at –20°C until use.

4. Etoposide stock: 10 mg of etoposide/ml dimethylsulfoxide (DMSO); store at –20°C until use.

### 2.6. Analysis of Redox forms of p53 and TRX In Vivo: Extraction From Cells in Presence of Iodoacetic Acid (IAA), Followed by Treatment with Iodoacetamide (IAM)

1. Urea buffer: 8 $M$ urea, 100 m$M$ Tris-HCl, pH 8.2; 1 m$M$ EDTA.

2. IAA stock solution: 600 m$M$ IAA in 1 $M$ Tris, pH 8.2.

3. IAM stock solution: 200 m$M$ IAM in 1 M Tris, pH 8.2.

4. Wash solutions: acetone:1 $N$ HCl (98:2, v/v), and acetone:1 $N$ HCl: H$_2$O (98:2:10, v/v), stored at 4°C.

5. 100 m$M$ dithiothreitol (DTT): make stock solution in ddH$_2$O and store at –20°C until use.

**2.7. Separation of Redox Forms of p53 and TRX on Urea-Poly-acrylamide Gels**

1. Acrylamide/bis-acrylamide stock solution: 30% acrylamide and 0.8% bis-acrylamide in $ddH_2O$, filter through Whatman #1 paper once dissolved (heat at 37°C with stirring if necessary to dissolve).

2. Separation (lower) gel stock solution: 1.5 $M$ Tris, pH 8.8.

3. Stacking (upper) gel stock solution: 0.5 $M$ Tris, pH 6.8.

4. Urea stock solution: 12.5 $M$ urea in $dH_2O$.

5. Separation gel contains 5.25%–9% acrylamide, 0.27% bis-acrylamide (w/v), 0.037 $M$ Tris, pH 8.8, and 8 $M$ urea (made from the 12.5 $M$ urea stock solution, preheated at 56°C just before use to get the urea into solution and then allowed to cool slightly before adding to gel mixture). For 10 mL of separation gel, add 67 µL of 10% ammonium persulfate (APS) and 20 µL of TEMED for polymerization of the acrylamide. For p53, use a final concentration of 5.25% acrylamide, whereas for TRX, use a final concentration of 9% acrylamide.

6. Stacking gel contains 2.5% acrylamide (w/v); 0.075% bis-acrylamide (w/v); 0.12 $M$ Tris-HCl, pH 6.8; and 8 $M$ urea. For 5 mL of stacking gel, add 34 µL of 10% APS and 10 µL of TEMED for polymerization of the acrylamide.

7. Urea sample buffer: 8 $M$ urea; 100 m$M$ Tris, pH 8.2; 1 m$M$ EDTA; 3.5 m$M$ DTT; 8% glycerol (v/v); and 0.1% bromophenol blue (w/v).

8. Running buffer: 25 m$M$ Tris, pH 8.3, and 192 m$M$ glycine.

9. Equilibration/transfer buffer: 25 m$M$ Tris and 192 m$M$ glycine, pH ~ 8.5 (no methanol present, but add if found to be necessary for transfer of specific protein).

10. Mini-gel electrophoresis and blotting apparatus (BioRad).

11. Immobilon-P membrane (0.45 µ), PVDF (Millipore).

**2.8. Immunodetection Protocol for the Visualization of Redox Forms of p53 and TRX With the Odyssey Infrared Scanner**

1. 1X PBS: 137 m$M$ NaCl, 2.7 m$M$ KCl, 4.3 m$M$ $Na_2HPO_4$, 1.4 m$M$ $KH_2PO_4$.

2. Odyssey blocking buffer (Li-Cor Biosciences).

3. 1X PBS + 0.1% Tween-20 (BioRad).

4. Primary antibodies: monoclonal mouse anti-human p53 (DakoCytomation clone DO-7) or monoclonal mouse anti-human TRX (BD Pharmingen, cat. no. 2G11).

5. IRDye800-conjugated affinity-purified goat α-mouse IgG, H&L (Rockland Inc., cat. no. 610-132-121)

6. Odyssey infrared scanner and Odyssey v1.1 software (Li-Cor Biotechnology).

**2.9. Cell Cycle Analysis of p53-Positive HCT116 Cells Versus p53-Negative HCT116 -/- Cells after Treatment With Reducing Agent WR1065**

1. HCT116 (p53-positive) and isogeneic HCT116−/− (p53-negative) human colon carcinoma cells, kindly provided by Dr. Burt Vogelstein.
2. McCoy's medium + 10% fetal bovine serum, 2 m$M$ glutamine, and penicillin/streptomycin.
3. CycleTEST™ PLUS DNA-staining kit (Becton Dickinson).
4. Becton Dickinson FACSCalibur flow cytometer.
5. Cell Quest and ModFit LT2.0 software (Becton Dickinson).

**2.10. Co-Transfection of H358 Cells With p53 +/− Thioredoxin to Study the Effect of a Reductant on p53-Induced Apoptosis**

1. NCI-H358 human bronchioloalveolar carcinoma cells (ATCC).
2. Complete medium: RPMI-1640 (Sigma) + 10% fetal calf serum + penicillin-streptomycin (final concentration: 100 units penicillin and 100 µg streptomycin/ml medium).
3. FuGene 6 transfection reagent (Roche).
4. DNAs for transfection: pcDNA3, wtp53pcDNA3, CMV-ADF (TRX) DNA, and carrier pBS.SK DNA.
5. G418 stock: make up stock at a concentration of 15 mg of G418/mL 0.1 $M$ HEPES, pH 7.4, and store at −20°C until use.
6. Stock Giemsa solution: add 48.2 mL of glycerol to 0.7 g of Giemsa powder, mix and place this solution in a 60°C oven for 2 h. After cooling to room temperature, add 46.2 mL of absolute methanol, mix, and filter through Whatman #1 paper. Store at room temperature.
7. Working Giemsa solution: dilute the Giemsa stock solution 1:25 in 0.01 $M$ Na$_2$HPO$_4$-NaH$_2$PO$_4$ buffer, pH 7.0.

# 3. Methods

**3.1. Purification of Wild-Type p53 Protein**

1. Grow Sf9 insect cells at 25°C in TNM-FH medium supplemented with 10% fetal calf serum until the cells reach 60–70% confluency in 100-mm plates. Remove the medium, and to each plate add 1.6 mL of a 1:1 dilution of high-titer NPVp53 recombinant baculovirus in TNM-FH medium plus 10% fetal calf serum. Rock the plates for 20 min at room temperature to allow for infection of the cells by the virus, and then add 9 mL of TNM-FH medium plus 10% fetal calf serum to each plate. After incubation at 25°C for approximately 65 h after infection, harvest the cells by vigorously pipetting the medium over the cells.

2. Transfer the cells and medium to a tube and centrifuge at 260 g for 5 min at 4°C to pellet the cells. Discard the supernatant, and wash the cells with ice-cold phosphate-buffered saline (PBS) and spin twice at 4°C as before. Resuspend the cell pellet in cold lysis buffer, 250 µl lysis buffer/cell pellet from the equivalent of one 100-mm plate, for 30 min on ice with pipetting up and down several times every 5–10 min. Then centrifuge the lysates at 18,000 rpm for 20 min at 4°C in a Beckman JA-20 rotor with microadaptors to obtain the supernatant for p53 purification.

3. Wash protein A-Sepharose beads chemically crosslinked to p53-specific monoclonal antibody PAb 421 (PAb421-PAS beads) three times with buffer containing 150 m$M$ Tris-HCl (pH 8.0), 150 m$M$ NaCl, 1 m$M$ EDTA, and 10% glycerol. Use approximately 75–100 µL of beads for each volume of lysate obtained from one 100-mm plate of infected cells. Mix the lysate and PAb421-PAS beads in a 1.5-mL microfuge tube for 15–16 h overnight by gentle inversion at 4°C. Then wash the beads four times with 0.75 mL/wash of buffer C, followed by three washes with 0.75 ml/wash of buffer D, and then three washes with 0.5 mL/wash of elution buffer (minus decapeptide).

4. Elute wild-type p53 protein by rotating the beads in 0.5 mL of fresh elution buffer with competing decapeptide KKGQSTSRHK-amide present at a final concentration of 0.1 mg/mL elution buffer, for 1 h at room temperature, followed by centrifugation to remove the beads. Store the purified protein in elution buffer at 4°C until used in the gel mobility shift assay.

5. Alternatively concentrate p53 in a Centricon-30 microconcentrator (Amicon) by centrifugation at 4°C followed by two washes of buffer containing 10 m$M$ HEPES (pH 7.8), 5 m$M$ KCl, 0.5 m$M$ MgCl$_2$, and 10% glycerol (*see* **Note** 1). Determine the protein concentration with the Bradford method, using bovine serum albumin as the standard.

6. This previously mentioned purification protocol uses solutions that have not been treated to remove divalent cations, and the p53 protein that is purified is referred to as "non-chelexed." To purify p53 that is free of exogenous divalent cations, all solutions used for purification of the protein must be pre-treated with Chelex-100 resin, simply by shaking each solution with a small amount of resin, centrifuging to remove the resin, and transferring the supernatant to a fresh tube. The p53 protein purified with the Chelex-treated solution is referred to as "chelexed," although the protein *per se* has not been exposed to Chelex.

**3.2. Preparation of $^{32}$P Labeled Target DNA**

1. For this assay, use the pBS.KS$_{Shay}$ plasmid that contains a p53 consensus sequence (5′-GGACATGCCCCGGGCAT-GTCC-3′) ligated into Bluescript KS at the *Bam*HI and *Pst*I sites. To prepare a $^{32}$P-labeled target DNA for the gel mobility shift assay, first digest the plasmid with *Bam*HI and *Pst*I to restrict out the DNA fragment containing the p53 consensus sequence. To 20 µg of pBS.KS$_{Shay}$ DNA, add 40 units each of *Bam*HI and *Pst*I restriction endonucleases and incubate in the appropriate 1x restriction buffer for 3 h at 37°C. At the end of the reaction, add 3x loading buffer with sodium dodecyl sulfate to a final concentration of 1X, and run the restricted DNA on a 1.5% agarose gel in 1X TAE buffer, in more than one lane if necessitated by volume of the sample. Turn off the power supply before the BPB dye front has migrated off the gel, since the Shay fragment runs just ahead of the dye.

2. Tombstone the Shay fragment containing the consensus sequence onto a piece of DEAE cellulose, by nicking the agarose on the anode side of the DNA, standing the small rectangular piece of NA45 DEAE cellulose (pretreated with the S & S protocol) up in the nicked agarose, and then running the DNA onto the DEAE cellulose at 100 V for 10 min. Elute the DNA off the DEAE cellulose by placing the pieces in sufficient high salt NET buffer in a large microfuge tube to keep them wet and then heating at 65°C for 30 min with several vortexes and quick spins during that time. Transfer the supernatant containing the DNA to a fresh microfuge tube, and to every 400 µl buffer containing DNA, add 4 µL of 1 *M* MgCl$_2$ and 800 µL of 200-proof ethanol to precipitate the DNA at –20°C for 1 h. Pellet the DNA in a microcentrifuge at 12,000 rpm at 4°C for 15 min, wash the pellet with cold 70% ethanol, spin again, air-dry the pellet, and then resuspend the DNA in 50 µL of distilled water.

3. Repreciptate the DNA to remove residual salt by adding 5 µL of 3 *M* Na-acetate plus 125 µL of 200-proof ethanol, mixing and leaving at –20°C for 30 min. Spin the DNA out at 12,000 rpm for 10 min at 4°C, wash with 200 µL of 95% ethanol, and respin. Finally, air-dry the pellet and resuspend in 25 µL of distilled water for use in the Klenow end-labeling reaction. To 25 µL of Shay DNA, add 3 µL of 10x Klenow buffer, 1 µl of a mix containing 5 m*M* each of dGTP, dCTP, and dTTP, 2 µL of α-$^{32}$P-dATP (3,000 Ci/mmol, 10 µCi/λ), and 2 units of Klenow DNA polymerase. Gently mix, quick spin, and then incubate at 37°C for 30 min with shielding. To ensure that the ends are completely filled in, chase the reaction by adding 1 µL of unlabeled 5 mM dATP to the

reaction, mix, quick-spin, and incubate at 37°C for an additional 5 min.

4. Finally, add 37 µL of TE to the reaction mix to bring the volume up to 70 µL and separate the unincorporated dNTPs from the labeled oligonucleotide Shay fragments by putting the reaction mix over a Nuctrap (Stratagene) push column. Prepare the column for use by adding 70 µL of 1X STE buffer to the top and pushing through with air from a syringe, just until a drop appears at the bottom of the column. After mounting the pre-wetted column in a column beta shield apparatus, apply the 70 µL of radioactive reaction mix and, with the syringe, push it slowly through the column into a large microfuge tube mounted in the beta shield base. Then push through another 70 µL of 1X STE to recover the rest of the labeled DNA in the eluate.

5. Estimate the volume of liquid eluted off of the column and precipitate the DNA by adding 1/10 volume 3 $M$ sodium acetate and 2–2.5 volumes of 200-proof ethanol, mixing and placing at –20°C for 1 h. Then spin in a microcentrifuge at 12,000 rpm for 15 min at 4°C, wash the pellet with 200 µL of 95% ethanol, spin again, remove the supernatant, and dry the pellet under gentle vacuum. Resuspend the $^{32}$P-labeled DNA in 20 µL of TE to be used in the gel mobility shift assay. Before use, estimate the concentration of the labeled DNA by resuspending a 2-µL aliquot in 2 µL of 15% Ficoll + 7 µL of dH$_2$O and running the sample against a given volume of low molecular weight Biomarker DNA (BioVentures, Inc.) on a 1.5% agarose gel in TAE with 3 µl stock EtBr solution for approximately 30 min at 100 V. Estimate the concentration of the Shay fragment by comparing the intensity with the 50 bp Biomarker fragment which is at a known concentration (10 ng µL$^{-1}$).

*3.3. Gel Mobility Shift Assay of p53 Binding to $^{32}$P-Labeled Target DNA*

1. Make up the mobility shift gel before conducting the p53-DNA binding assay. For a 4% polyacrylamide mobility-shift gel, combine 20 mL of low-ionic-strength gel mix, 50 µL of 30% APS, and 17 µL of TEMED. Pipet the mixture between the glass plates, one of which has been siliconized with Rain-x. Quickly insert the comb into the single-phase gel and allow the gel to polymerize at least 20 min before moving it into the cold room.

2. The gel mobility shift assay used is similar to that described by Hupp et al. *(9)*, with minor variations. Pre mix $^{32}$P-end-labeled Shay DNA with Bluescript SK$^+$ DNA at a ratio of 1:25 (typically 2 ng of Shay DNA: 50 ng BS.SK$^+$ DNA per microliter). When DTT is present in the reaction mix, be sure to preincubate the requisite amount of purified p53

protein (50–900 ng) with 5 m$M$ DTT in DNA binding buffer for 20 min at room temperature before the addition of DNA or TE (in those samples where DNA is not present). Simultaneously incubate samples containing p53 protein but no DTT with an equivalent volume of H$_2$O before the addition of DNA, so that the reaction volumes are similar. Then, add 2 µL of the DNA mixture to 350 ng of purified wild-type p53, prepared with either chelexed (C) or non-chelexed (NC) solutions, in a final volume of 20 µL and incubate at room temperature for 30 min.

3. During incubation of the DNA-binding assay, electrophorese the pre-chilled low-ionic-strength 4% polyacrlamide gel at 100 V in cold running buffer (0.33X TBE and 0.1% Triton X-100) at 4°C. Stop the gel and wash out each well with running buffer just prior to loading the samples. At the end of the incubation period, carefully load 10 µL of each sample onto the gel while the gel is running at 100 V and electrophorese at 200 V for 1 h at 4°C.

4. Once electrophoresis is finished, remove the gel carefully from the glass plate, in preparation for blotting, by laying one dry piece of precut Whatman 3-mm filter paper onto the gel, turning the gel + plate over, and carefully lifting the plate away from the gel + filter paper, with the gel adhering to the filter paper.

### 3.4. Blotting and Immunodetection of Mobility-Shifted p53-DNA Complexes

1. Set up the blotting apparatus and membrane while the gel is running. Thoroughly soak two foam pads for the blotting apparatus in 0.33X strength TBE buffer. Cut four pieces of Whatmann 3-mm filter paper to a size slightly larger than the trimmed gel. Soak three of these pieces in 0.33x TBE, but keep one piece dry (used to remove the gel from the plate). Cut one piece of Nytran plus membrane (*see* **Note** 2) to size, soak in dH$_2$O for 15 min, and then briefly soak in 0.33X TBE just before setting up the blot.

2. Assemble the sandwich for blotting in the following sequence on the blotting cassette: one wet foam pad, one piece of wet filter paper, filter paper with gel adhering to it on top (then 0.33X TBE to wet the gel), wet Nytran plus membrane, two wet pieces of filter paper. Roll a glass pipet over the sandwich in both directions, pressing down firmly to squeeze out any bubbles. Then place the second wet foam pad on top of the sandwich, close the cassette and put into the blotting apparatus with buffer present.

3. Blot the gel overnight at 12 V onto Nytran plus membrane in 0.33X TBE (without Triton X-100) at room temperature. Immediately after disassembly of the blot, crosslink the DNA onto the membrane in the Stratalinker using the Auto

setting (1200 µJ). Then, rinse the membrane with distilled water, air-dry thoroughly, and expose to Kodak XAR-5 film, with or without an enhancing screen, for the requisite amount of time needed to detect the ³²P-labeled DNA in the mobility-shifted p53-DNA complexes. After detection of the radiolabeled DNA by autoradiography, visualize p53 protein on the same Nytran plus membrane (*see* **Note** 3) by immunodetection as follows. An example of the results obtained is shown in **Fig. 9.1**.

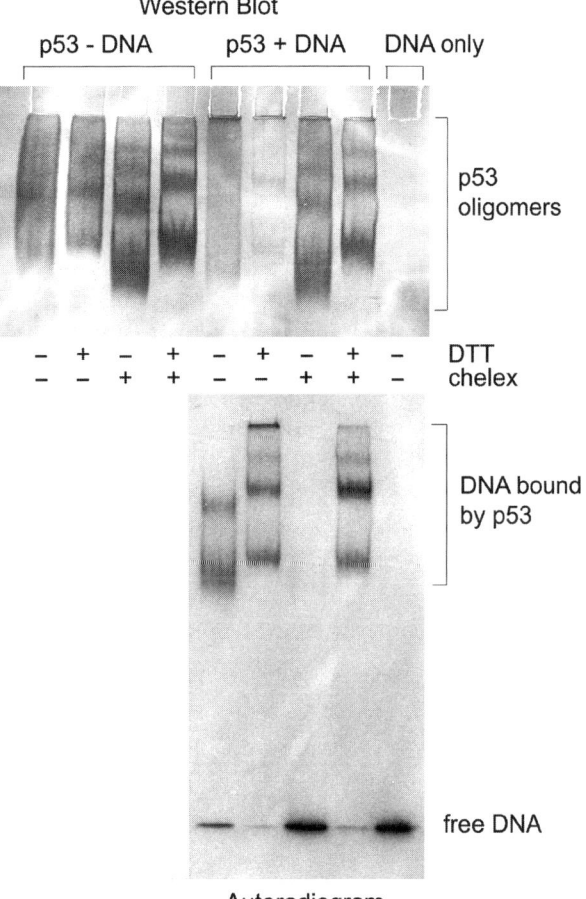

Fig. 9.1. Binding of chelexed vs nonchelexed p53 to consensus DNA. 350 ng of each p53 was used in a gel mobility shift assay, after pretreatment with 5 mM DTT where indicated, prior to the addition of target DNA. The p53 protein is shown in the Western blot in the *upper panel*, whereas the radioactive DNA when present is shown in the corresponding lanes of the autoradiogram in the *lower panel*. Chelexed p53 binds target DNA only when reduced and does not bind DNA in the oxidized state. In contrast, oxidized nonchelexed p53 binds target DNA almost as efficiently as reduced nonchelexed p53. Oxidized protein–DNA complexes (–DTT) migrate faster than reduced protein–DNA complexes (+DTT). Likewise, oxidized p53 protein oligomers (–DTT) migrate faster than reduced p53 protein oligomers (+DTT).

4. After wetting the membrane in distilled water, rinse it for 30 min in Tris-buffered saline (TBS) and then gently rock it for 1 h in blocking buffer at room temperature. Wash the membrane twice for 15 min each with TBST. To detect p53, incubate the membrane for 2 h with gentle rotation in a mixture of monoclonal antibodies PAb 421 and PAb 248, each at a 1:500 dilution in TBST, and PAb 242 at a 1:250 dilution in TBST. After incubation, remove unabsorbed primary antibodies, and wash the membrane twice for 15 min each with TBST.

5. Add secondary antibody, biotinylated goat anti-mouse IgG, to fresh TBST at a 1:667 dilution for 30 min with gentle rotation, followed by two 15-min washes in TBST to remove unabsorbed secondary antibody. Then. add streptavidin-conjugated alkaline phosphatase to fresh TBST at a 1:10,000 dilution and incubate for 30 min with gentle rotation. Follow this incubation with two 15-min washes with TBS (no Tween-20 present) and then one 15-min wash with substrate buffer.

6. Just before using, mix 35 mL fresh substrate buffer with 154 µL of nitroblue tetrazolium chloride and 116 µL of 5-bromo-4-chloro-3-indolylphosphate p-toluidine salt, then add to the membrane and rotate at room temperature in a container wrapped with aluminum foil, to avoid exposure to light. Allow the purple color to develop until the desired signal is achieved, about 45–60 min later, then wash the membrane with several changes of water and allow it to air dry.

### 3.5. Effect of Redox Perturbing Agents, Such as Aminothiol WR1065, Diamide, or Etoposide, on the Redox State of p53 and TRX in Cells Treated With These Agents

1. To analyze the effect of a particular agent on the redox state of p53, treat actively growing HCT116 cells that are approximately 30–40% confluent with the specific redox-perturbing agent at varying concentrations for varying periods of time prior to harvesting the cells as described below in **Section** 3.6. If testing the effect of the reducing agent aminothiol WR1065 on cells, pretreat the cells with aminoguanidine (AG) at a final concentration of 4 m$M$ for 10 min before the addition of WR1065, to prevent the degradation of WR1065 into cytotoxic metabolites by intracellular Cu-dependent amine oxidases *(10)*. Then add WR1065 to a final concentration of 1 m$M$ and incubate the cells overnight in a $CO_2$ incubator at 37°C for 16–18 h before harvest.

2. To treat HCT116 cells with etoposide, add 5 µL of etoposide stock to 10 mL of medium to give a final concentration of 5 µg of etoposide per milliliter of medium, rock well to mix, and incubate the cells overnight in a $CO_2$ incubator at 37°C for the requisite period of time prior to harvest. Note that overnight treatment with etoposide results in a flattening

and enlarged appearance of the cells, with much more distinct nuclei and nucleoli.

3. For cells to be treated with diamide, add the appropriate amount of stock solution to cells to give a final concentration of 10 m$M$ and incubate for only 30 min before harvest, because cells tend to come off the plate.

4. If treatment with the redox-perturbing agent causes the cells to detach from the plate, as in the case of diamide, then it is necessary to modify the harvest procedure. Pipet the cells into a centrifuge tube on ice, spin at 260 g for 3 min at 4°C, discard the supernatant, wash the cell pellet once with cold PBS, spin again, and remove the supernatant. Then add the equivalent of 200 μl urea lysis buffer containing IAA per 10 cm plate to the cell pellet, resuspend, and proceed as delineated in Sect. 3.6 below.

***3.6. Analysis of Redox Forms of p53 and TRX In Vivo: Extraction From Cells in the Presence of IAA, Followed by Treatment with IAM***

1. The method for the analysis of the redox forms of p53 is derived from the methodology developed by Takahashi and Hirose *(11)* and by Bersani et al. *(12)* for analysis of the redox state of thioredoxin in cells. Before harvest of the cells, suction off the medium and wash the cells 2X with 5 mL of 1X PBS/wash, and place the plate on ice in preparation for harvest. For cell lysis and carboxymethylation of free sulfhydryl groups, pipet 200 μL of urea buffer containing 30 m$M$ IAA onto each 10-cm plate of cells and rock to cover the monolayer. An appropriate amount of 600 m$M$ IAA stock solution (prepared in 1 $M$ Tris, pH 8.2) is added to urea buffer to give a final concentration of 30 m$M$ IAA. After approx 5 min on ice, scrape the extract into one corner of the plate and transfer with a micropipettor to a chilled 1.5-mL microfuge tube.

2. After harvesting all the plates, incubate the tubes with extracts at 37°C for 15 min, and then clarify the samples by microcentrifugation for 2 min at 17,900 g at 4°C. Transfer each supernatant into two tubes with approximately 200 μL in each tube.

3. Remove the unreacted IAA by adding 1.5 mL of cold acetone-1 $N$ HCl (98:2 v/v) to precipitate the proteins for approximately 5 min on ice, and then microcentrifuge at 11,000$g$ for 5 min at 4°C. Wash the pellet 3X by resuspension in cold acetone-1 $N$ HCl-H$_2$O (98:2:10 v/v/v), followed by centrifugation as above. Pool the two separate pellets from each sample after the second wash.

4. For reduction of disulfides with DTT, resuspend the final acetone precipitate in 95 μL of urea buffer containing 3.5 m$M$ DTT and incubate the samples for 30 min at 37°C. Reduction

with DTT generates free sulfhydryl groups which are then alkylated with IAM. To each sample, add an appropriate amount of 200 m$M$ IAM stock solution (prepared in 1 $M$ Tris, pH 8.2) to give a final concentration of 10 m$M$ IAM. Mix, incubate the samples in IAM for 15 min at 37°C, and then microcentrifuge at 11,000 g for 1 min at 4°C, and transfer the supernatant to a fresh microfuge tube.

5. Freeze the samples at –70°C until use (*see* **Note** 4), at which time thaw and determine the protein concentrations using the Bradford method.

### 3.7. Separation of Redox Forms of p53 and TRX on Urea-Polyacrylamide Gels

1. Depending on the protein being analyzed, make up a urea-polyacrylamide separation gel containing anywhere from 5.25% acrylamide (for p53) to 9% acrylamide (for TRX; *see* **Note** 5). Make up samples containing 20 µg of cell extract in urea sample buffer and heat at 37°C for approx 15 min before loading. After removing the comb from the stacking gel, wash out the wells with running buffer immediately before loading the samples. Run a mini-gel (~7 × 10 cm in size) at 5 mA constant current for approximately 1.5 h, until the blue dye just reaches the bottom of the gel.

2. In the meantime, prepare the blotting apparatus and make up equilibration/transfer buffer containing 25 m$M$ Tris, pH ~ 8.5, and 192 m$M$ glycine (without methanol). Cut a piece of Immobilon-P membrane appropriate for the size of the gel to be transferred. Pre-wet the membrane with 100% methanol and then equilibrate the membrane in transfer buffer. Pre-soak the sponges in transfer buffer until they are thoroughly saturated. Remove the gel from the gel apparatus and equilibrate in transfer buffer with gentle shaking for 15–20 min at room temperature.

3. Assemble the blot with the membrane towards the anode, and place the assembly frame into transfer buffer in the mini-blotting apparatus. Transfer the proteins from the gel onto the membrane at 12 V constant voltage overnight at room temperature. Then disassemble the blot, nick a corner of the blot to maintain orientation, and place in 1X PBS in preparation for immunodetection of p53 or any other protein being studied.

### 3.8. Immunodetection Protocol for the Visualization of Redox Forms of p53 and TRX with the Odyssey Infrared Scanner

1. After rinsing the blot for 10 min in 1X PBS, add enough Odyssey blocking buffer to cover the membrane and gently rock for 1 h at room temperature to block nonspecific binding sites. Remove the blocking buffer (can be reused), and add a similar volume of blocking buffer + 0.1% tween-20 + the primary antibody at a given dilution, ranging from 1:1,000 to 1:1,500.

2. For detection of p53, use monoclonal DO-7 mouse anti-human p53 at a dilution of 1:1,000; for detection of TRX, use monoclonal mouse anti-human TRX at a dilution of 1:1,500. Incubate the blot with the primary antibody with gentle shaking for 1 h at room temperature, making sure that the membrane remains moist. After incubation, remove the Odyssey buffer containing the primary antibody and store at 4°C if it is to be used again.

3. Wash the blot four times with PBS + 0.1% Tween-20 for 5 min each time at room temperature with gentle shaking.

4. Just before use, dilute the secondary antibody, IRDye 800-conjugated goat α-mouse IgG, 1:10,000 in blocking buffer + 0.1% Tween-20, and keep dark. Add to the blot and gently shake for 30 min at room temperature in a container wrapped in aluminum foil to avoid exposure to light. After incubation, remove the solution and store at 4°C if it is to be used again.

5. Wash the blot as before, four times with PBS + 0.1% Tween-20 for 5 min each time at room temperature with gentle shaking, in the container wrapped in foil. Then rinse 1 time in 1X PBS (without detergent), store the blot wet in 1X PBS, and scan the moist blot on the Odyssey infrared scanner at a wavelength of 800 nm. The redox forms of p53 and TRX in cells treated with redox agents can be seen in **Fig. 9.2**.

6. Alternatively, immunodetection of the Western blot can be performed with a secondary antibody coupled to horseradish peroxidase, binding of which is detected with the enhanced chemiluminescence detection kit (*see* **Note** 6).

***3.9. Cell Cycle Analysis of p53-Positive HCT116 Cells Versus p53-Negative HCT116-/- Cells After Treatment With Reducing Agent WR1065***

1. Subculture and set up HCT116 cells or HCT116-/- cells in 10-cm Petri dishes at a cell density of $2.25 \times 10^6$ cells/plate. Treat the cells the following day with a redox-perturbing agent, in this case WR1065, when the cells are approximately 30–40% confluent. Pretreat the cells with aminoguanidine (AG) at a final concentration of 4 m$M$ for 10–15 min, and then add the aminothiol WR1065 to a final concentration of 1 m$M$. Incubate the cells overnight in a $CO_2$ incubator at 37°C for a specified time period prior to harvest.

2. After observing the cell morphology and estimating the confluency, trypsinize the cells and transfer the cell suspension into medium with serum in order to inactivate the trypsin. Centrifuge for 5 min at 1,200 rpm at 4°C, discard the supernatant, and wash the cell pellet with cold 1X PBS, centrifuging as before. Resuspend the cell pellet harvested from one plate of cells in 0.5 mL of the citrate/sucrose/DMSO

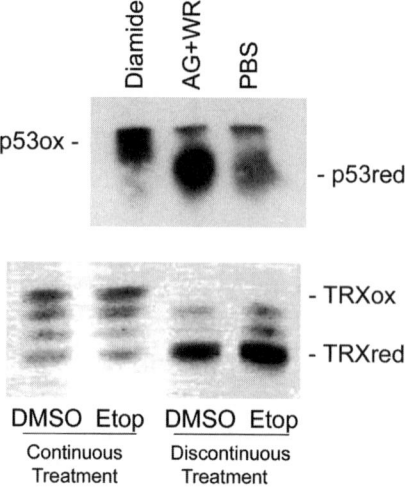

Fig. 9.2. Analysis of the in vivo redox state of p53 and TRX in cells treated with redox agents, with protein extraction from cells in the presence of IAA, followed by treatment of the extracts with IAM. Protein samples (15 µg) are separated by urea-polyacrylamide gel electrophoresis, and the different redox forms of the proteins are then immunodetected with antibody specific for p53 (*upper panel*) or TRX (*lower panel*). In the upper panel, treatment of HCT116 cells with WR1065 results in a marked increase in the amount of reduced p53 relative to the control PBS treatment; diamide treatment results in p53 being converted to an oxidized, slower-migrating form. In the lower panel, there is no obvious difference seen in the relative levels of oxidized versus reduced TRX in HCT116 cells treated with etoposide (Etop) as opposed to DMSO in the control. Differences are found in the array of redox forms of TRX detected, depending on whether the IAA-IAM treatment is continuous or discontinuous (refer to Note 4), so it is important to standardize the treatment across samples for correct interpretation of the results. Treatment of cells with diamide results in the conversion of these three to four visible forms to two forms, which migrate more slowly (data not shown).

solution provided in the CycleTEST™ PLUS DNA-staining kit. Proceed to stain the cells for immediate analysis, or alternatively freeze them down at –80°C for later analysis. If frozen down, then at the time of analysis thaw the cells quickly at 37°C and, once thawed, place the tube on ice.

3. If it is desired to analyze the proteins in the same samples that will be analyzed for relative distribution in the cell cycle, then divide each cell sample into two aliquots when resuspended in the cold 1x PBS wash in **step 2**. After centrifugation, process this cell pellet for protein analysis as delineated in **Note** 7.

4. For cell cycle analysis, pipette 100 L or $5 \times 10^5$ of thawed cells (*see* **step 2** in this section) into a fresh tube, centrifuge for 5 min at 400$g$ at room temperature, and remove the supernatant completely. Resuspend the pellet in 120 µL

of solution A, containing trypsin in a spermine tetrahydro-chloride detergent buffer, and gently mix by tapping the tube by hand, with no vortexing. Incubate 10 min at room temperature. Then add 100 μL of solution B, containing trypsin inhibitor and ribonuclease A, and again mix by hand and incubate 10 min at room temperature. Do not remove solutions A or B. Proceed to add 100 μL of solution C containing propidium iodide and spermine tetrahydrochloride in citrate stabilizing buffer to each tube. Mix by hand, and incubate 10 min at 4°C in the dark.

5. Keep the cell samples on ice in the dark until diluting an aliquot of the sample with FACSFlow and acquiring the cell cycle distribution data with a FACSCalibur flow cytometer, according to the instructions of the manufacturer.

6. Analyze the data with Cell Quest and ModFit LT2.0 soft-ware to determine the percentage of cells in the $G_1$, S, and $G_2$ phases of the cell cycle as well as the percentage of sub-diploid cells in each sample. A plot of the percentage of cells in each phase of the cell cycle +/– treatment with WR1065 (**Fig. 9.3**) shows a marked increase in G1 cell cycle arrest in the p53-positive HCT116 cells in response to the reduct-ant WR1065 in contrast to a weak G1 arrest response to WR1065 in the p53-negative HCT116-/- cells, in agree-ment with the data presented in Mann and Hainaut *(13)*.

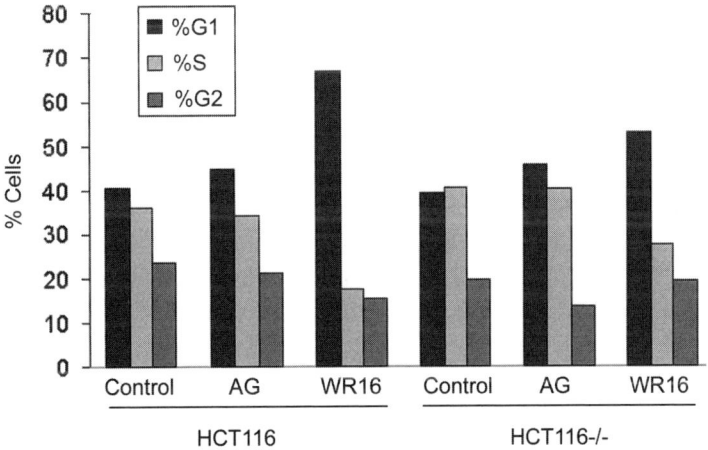

Fig. 9.3. Effect of WR1065 treatment on the cell cycle distribution of HCT116 (p53-positive) versus HCT116-/– (p53-negative) cells. Both cell lines were treated for 16 h with either PBS (Control), aminoguanidine (AG), or AG + WR1065 (WR16), and then harvested for FACSCalibur flow cytometric analysis. The percentages of HCT116 and HCT116-/- cells in the G1, S, and G2 phases of the cell cycle, as determined using Cell Quest and ModFit LT2.0 software analysis of the FACS data, have been plotted for each of the treatment protocols.

***3.10. Co-Transfection
of H358 Cells With p53
+/– Thioredoxin to
Study the Effect of a
Reductant on p53-
Induced Apoptosis***

1. Subculture H358 cells into 5-cm Petri dishes and transfect the next day, when approximately 30–40% confluent. Reduce the volume of the original medium (RPMI-1640 + 10% fetal calf serum + penicillin-streptomycin) on each plate to 2 mL/plate just before transfection. Transfect with a total of 2 µg DNA/plate in the presence of FuGene 6 transfection reagent. Premix the test DNA mixtures (0.5 µg total) with 1.5 µg of pBS.SK (Stratagene) carrier DNA, and bring to a final volume of 10 µL with TE. The test DNAs are all contained within the pcDNA3 vector which confers gentamycin (G418)-resistance to those cells that are successfully transfected. Transfect the plates with either 0.5 µg of pcDNA3 vector alone, 0.25 µg of human wtp53pcDNA3 + 0.25 µg pcDNA3, 0.25 µg of CMV-ADF (TRX) + 0.25 µg of pcDNA3, or 0.25 µg of CMV-ADF (TRX) + 0.25 µg of wtp53pcDNA3.

2. For each transfection mixture, add 94 µL of RPMI-1640 medium minus serum to a 0.5-mL microfuge tube, and pipet 6 µL of FuGene 6 reagent directly into the medium so as not to make contact with the walls of the tube. Tap each tube or flick to mix, and leave at room temperature for 5 min. Then, pipet the FuGene/medium mixture drop-wise onto a 10-µL aliquot of premixed DNAs, tapping the tube to mix, and leave at room temperature for a minimum of 15 min but not more than 30 min. Add the final transfection reagent/DNA complex to each respective plate of cells in a drop-wise manner and rock the plate immediately to distribute the mixture evenly throughout the medium. Incubate the plates in a 37°C incubator with 5% $CO_2$.

3. The following day, remove the medium from each plate, wash with 2.5 mL of RPMI-1640 minus serum, and add 4 mL of complete medium with a concentration of G418 sufficient to select for the presence of the transfected DNA conferring resistance (for H358 cells, 750 µg G418/mL medium is sufficient). Every 3–4 d, suction off the medium containing dead cells plus cell debris, and add 4 mL of fresh complete medium + G418 to each plate.

4. Approx 10–14 d after transfection, harvest and stain the plates for G418-resistant colonies of cells as follows. Suction off the medium, wash each plate two times with 2.5 mL of 1X PBS or 1X TD, and then add 2.5 mL of 200-proof ethanol to each plate for fixation of the cells for 5–10 min at room temperature. Suction off the ethanol, and allow the plates to air dry before proceeding with Giemsa staining.

5. Just before use, dilute the Giemsa stock solution 1:25 in 0.01 $M$ $Na_2HPO_4$-$NaH_2PO_4$ buffer, pH 7.0, to make a working Giemsa solution. Add 2.5 mL of the working Giemsa solution

H358 + vector            H358 + TRX

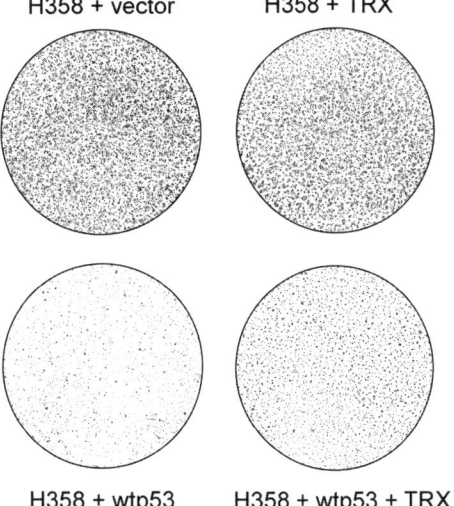

H358 + wtp53        H358 + wtp53 + TRX

Fig. 9.4. Analysis of apoptosis in p53-negative H358 cells co-transfected with wild-type p53 +/– reductant thioredoxin (TRX). Permanent transfectants were selected for resistance to G418, and 12 days later G418-resistant clones were stained with Giemsa. Comparison of the lower plates in the figure shows that TRX partially protects cells from induction of apoptosis by p53. Relative mean optical densities measured by Scion Image analysis of scanned images of the transfected H358 cells are shown in parentheses: + vector (34.72), + TRX (28.19), + wtp53 (4.01), + wtp53 + TRX (13.30) DNAs.

to each plate, rock to cover the plate completely, and leave at room temperature for 10–15 min. Then suck off the stain and immediately dip each plate in two changes of water to remove the excess stain, and allow to air dry.

6. Quantitate the relative mean optical densities by Scion Image for Windows (*see* **Note** 8) analysis of scanned images of the plates containing the G418-resistant H358 cells. Scanned plates from a typical transfection experiment are shown in **Fig. 9.4**, with their relative densities from the image analysis listed in the figure legend. The results suggest that the presence of TRX decreases the induction of apoptosis by wtp53 in H358 cells.

## 4. Notes

1. If purified p53 is to be concentrated in a Centricon-30 microconcentrator, recovery of the protein can be enhanced fivefold to sevenfold by using solutions that have been pre-treated with Chelex-100 from the time of lysis of the infected Sf9 cells and throughout the purification procedure.

2. Testing of other membranes both non-nylon (nitrocellulose, DEAE cellulose) and nylon (duralon, nytran) has shown that Nytran plus is far more effective in its retention of DNA 50–60 nucleotides in length.

3. If desired, it is possible to do a double blot in which the mobility shift gel is assembled with a piece of nitrocellulose BA-85 (Schleicher & Schuell, pore size 0.45 μ) immediately on top of the gel, with a piece of Nytran plus membrane on top of the nitrocellulose membrane. In this case, pre-wet the nitrocellulose in 0.33X TBE for 15 min before assembling the blot. After blotting, wash the nitrocellulose with several changes of $dH_2O$ before immunodetection. It is also possible to store the membrane dry at this point if time is limited, and re-wet with $dH_2O$ later prior to proceeding with immunodetection.

4. For reproducible results, it is important to complete the full reaction of the proteins with IAA followed by IAM before freezing down the samples at –70°C, rather than discontinuous treatment where the samples are frozen down after treatment with IAA and then thawed to treat with IAM at a later point in time. The artefactual differences can be seen in the lower panel of **Fig.9.2**.

5. Migration of a specific protein in a urea-polyacrylamide gel cannot always be predicted on the basis of the molecular weight of the protein. For example, cytochrome *c* with a molecular weight similar to that of TRX actually migrates at a rate similar to that of p53, so it should be run on a 5.25% gel, rather than a 9% gel, to separate its redox forms.

6. For ECL detection of the Western, incubate the membrane in PBS + 0.05% NP-40 + 5% nonfat dry milk for 1 h at room temperature with gentle shaking. Then add the primary antibody diluted in PBS + 0.05% NP-40 + 0.5% powdered milk at the same concentration as specified for the Odyssey detection, and incubate for 1 h at room temperature with shaking. Wash the membrane four times with PBS + 0.05% NP-40 for 15 min/wash. Incubate the membrane with the secondary antibody, horseradish peroxidase-coupled goat anti-mouse IgG diluted 1:3,000 in PBS + 0.05% NP-40, for 1 h at room temperature: Remove unbound secondary antibody by washing the membrane four times with PBS + 0.05% NP-40 for 15 min/wash. To detect the bound HRP-coupled antibody, mix the two ECL reagents 1:1 just before use, place the membrane on a piece of Saran wrap protein side up and pipette on the mixed reagent so that the membrane is fully covered. Incubate for 1 min at room temperature, and then carefully lift the membrane with forceps to drain off the excess reagent and wrap the membrane in Saran

wrap. Expose the wrapped membrane to a sheet of Amersham Hyperfilm-ECL in a film cassette for different periods of time, as judged by an initial exposure of 1 min.

7. For protein sodium dodecyl sulfate -polyacrylamide gel analysis, remove the 1X PBS from the centrifuged cells and resuspend the cell pellet in 200 μL of EMSA buffer A (20 m$M$ HEPES, pH 7.6; 20% glycerol; 10 m$M$ NaCl; 1.5 m$M$ MgCl$_2$; 0.2 m$M$ EDTA; 1 m$M$ DTT; and 0.1%NP-40) with protease inhibitors (0.5 m$M$ PMSF, 0.5 μg/mL leupeptin, 2 μg/mL aprotinin, and 0.7 μg/mL pepstatin A), until no clumps are present. Transfer to a large microfuge tube and leave on ice for 10 min. Then centrifuge at 5,000$g$ for 4 min at 4°C in a refrigerated microfuge, and transfer the cytoplasmic extract to a clean tube for storage at –70°C. Add 150 μL of EMSA buffer B (same as EMSA buffer A, but with 500 m$M$ NaCl final concentration) with protease inhibitors (same as above) to each pellet, resuspend the pellet thoroughly, and leave on ice for 30 min. Centrifuge at 14,000 rpm for 15 min at 4°C, and carefully transfer the nuclear extract to a fresh tube and store at –70°C until analysis on sodium doceyl sulfate-polyacrylamide gels.

8. Scion Image for Windows (Scion Corporation) can be downloaded at no charge to the user. It is the Windows version of Scion Image and is similar to NIH Image, originally written at the National Institutes of Health for the Macintosh. NIH Image (1.62) is a public domain image processing and analysis program for the Macintosh that can also be downloaded for free.

## Acknowledgments

The author would like to thank Robin Rainwater and Dorothy Parks for their significant contributions (3, 14) to **Fig.** 9.1 of this chapter.

## References

1. Hainaut, P. and Mann, K. (2001). Zinc binding and redox control of p53 structure and function. *Antioxid Redox Signal* 3, 611–623.
2. Hainaut, P. and Milner, J. (1993). Redox modulation of p53 conformation and sequence-specific DNA binding in vitro. *Cancer Res* 53, 4469–4473.
3. Rainwater, R., Parks, D., Anderson, M. E., Tegtmeyer, P., and Mann, K. (1995). Role of cysteine residues in regulation of p53 function. *Mol Cell Biol* 15, 3892–3903.
4. Cho, Y., Gorina, S., Jeffrey, P. D., and Pavletich, N. P. (1994). Crystal structure of a p53 tumor suppressor DNA complex: understanding tumorigenic mutations. *Science* 265, 346–355.

5. Powis, G. and Montfort, W. R. (2001). Properties and biological activities of thioredoxins. *Annu Rev Pharmacol Toxicol* **41**, 261–295.

6. Stenger, J. E., Mayr, G. A., Mann, K., and Tegtmeyer, P. (1992). Formation of stable p53 homotetramers and multiples of tetramers. *Mol Carcinogenesis* **5**, 102–106.

7. Simanis, V. and Lane, D. P. (1985). An immunoaffinity purification procedure for SV40 large T antigen. *Virology* **144**, 88–100.

8. Funk, W. D., Pak, D. T., Karas, R. H., Wright, W. E., and Shay, J. W. (1992). A transcriptionally active DNA-binding site for human p53 protein complexes. *Mol Cell Bio.* **12**, 2866–2871.

9. Hupp, T. R., Meek, D. W., Midgley, C. A., and Lane, D. P. (1992). Regulation of the specific DNA binding function of p53. *Cell* **71**, 875–886.

10. Meier, T. and Issels, R. D. (1995). Degradation of 2-(3-aminopropylamino)-ethanethiol (WR-1065) by Cu-dependent amine oxidases and influence on glutathione status of Chinese hamster ovary cells. *Biochem Pharmacol* **50**, 489–496.

11. Takahashi, N. and Hirose, M. (1990). Determination of sulfhydryl groups and disulfide bonds in a protein by polyacrylamide gel electrophoresis. *Anal Biochem* **188**, 359–365.

12. Bersani, N. A., Merwin, J. R., Lopez, N. I., Pearson, G. D., and Merrill, G. F. (2002). Protein electrophoretic mobility shift assay to monitor redox state of thioredoxin in cells. *Methods Enzymol.* **347**, 317–326.

13. Mann, K. and Hainaut, P. (2005). Aminothiol WR1065 induces differential gene expression in the presence of wild-type p53. *Oncogene* **24**, 3964–3975.

14. Parks, D., Bolinger, R., and Mann, K. (1997). Redox state regulates binding of p53 to sequence-specific DNA, but not to non-specific or mismatched DNA. *Nucl Acids Res* **25**, 1289–1295.

# Chapter 10

# Redox Regulation and Trapping Sulfenic Acid in the Peroxide-Sensitive Human Mitochondrial Branched Chain Aminotransferase

## Susan M. Hutson, Leslie B. Poole, Steven Coles, and Myra E. Conway

## Abstract

The human branched chain aminotransferase enzymes are key regulators of glutamate metabolism in the brain and are among a growing number of redox-sensitive proteins. Studies that use thiol-specific reagents and electrospray ionization mass spectrometry demonstrate that the mitochondrial BCAT enzyme has a redox-active CXXC center, which on oxidation forms a disulfide bond (RSSR), via a cysteine sulfenic acid intermediate. Mechanistic details of this redox regulation were revealed by the use of mass spectrometry and dimedone modification. We discovered that the thiol group at position C315 of the CXXC motif acts a redox sensor, whereas the thiol group at position C318 permits reversible regulation by forming an intrasubunit disulphide bond. Because of their roles in redox regulation and catalysis, there is a growing interest in cysteine sulphenic acids. Therefore, development of chemical tags/methods to trap these transient intermediates is of immense importance.

**Keywords:** CXXC motif, oxidation, sulphenic acid, dimedone, mass spectrometry.

## 1. Introduction

Redox imbalance in the cell, as a result of oxidative stress, has been implicated in the pathogenesis of several diseases, including neurodegenerative disorders and cardiovascular disease (reviewed in **ref**. *1*). Yet, the molecular details of these redox mechanisms are poorly understood *(1)*. Although cysteine residues are one of the least abundant amino acid residues in proteins, their thiol(ate) groups are potent nucleophiles, suggesting they play a fundamental in redox biochemistry *(2,3)*. An exceptionally reactive and often transient intermediate of

John T. Hancock (ed.), *Methods in Molecular Biology, Redox-Mediated Signal Transduction, vol. 476*
© 2008 Humana Press, a part of Springer Science + Business Media, Totowa, NJ
DOI: 10.1007/978-1-59745-129-1_10

cysteine oxidation, cysteine sulfenic acid (Cys-SOH), is of biological significance *(3,4)*. Roles for sulfenic acid range in diversity from the redox regulation of the bacterial transcriptional activator (OxyR) of *Escherichia coli (5,6)*, to the redox control of protein tyrosine phosphatases *(7–9)*.

Several tools for trapping Cys-SOH include 5,5-dimethyl-1,3-cyclohexanedione, (dimedone) and 7-chloro-4-nitrobenz-2-oxa-1,3-diazole (NBD-Cl) *(10–12)*. Although dimedone is a highly specific reagent, reacting only with sulfenic acid, it cannot be characterized spectrally. Nevertheless, the dimedone adduct can be detected with the use of electrospray ionization mass spectrometry (ESI-MS), resulting in an increase of 138 ± 5 Da. Unlike dimedone, NBD-Cl can be detected spectrally at 420 nm after reaction with a reduced cysteine thiol/ate and at 347 nm after reaction with a cysteine sulphenic acid *(3,11)*. Recently, several advances in the development of chemical probes to target Cys-SOH will prove pivotal in the future analysis and identification of these interactions in redox sensitive proteins *(13,14)*.

The mammalian BCAT proteins are among a large number of redox-sensitive proteins that undergo thiol modification *(15)*. Evidence for redox cycling in mitochondrial hBCAT (hBCATm) was obtained with ESI-MS of a 5, 5′- dithiobis(2-nitrobenzoic acid) (DTNB)-labeled protein, which showed a mixture of labeled species in the sample rather than the expected mass of the protein with two TNB molecules, as dictated by DTNB titration *(15)*. Subsequent oxidation studies with WT and mutant hBCATm proteins using dimedone and ESI-MS showed that the thiol group at position C318 effectively competed with a second DTNB reacting with the protein and instead led to loss of TNB and disulfide bond formation between C315 and C318 after DTNB treatment. The presence of C318 also prevented the overoxidation of the cysteine sulfenic acid at position C315 and, thus, irreversible inactivation of the enzyme *(12)*.

## 2. Materials

### 2.1. Purification of hBCATm and the CXXC Mutants

1. Extraction buffer A: 0.01 *M* Tris-HCl, pH 8.0; 0.1 *M* sodium phosphate; and 5 m*M*

2. β-mercaptoethanol. Prepare fresh and store at 4°C.

3. Extraction buffer AU: 0.01 *M* Tris-HCl, pH 8.0; 0.1 *M* sodium phosphate; 5 m*M* β-mercaptoethanol, and 4 *M* urea. Prepare fresh and store at 4°C.

4. Nickel-NTA resin elution Buffer A: 0.01 *M* Tris-HCl, pH 8.0; 0.1 *M* sodium phosphate; and 5 m*M* β-mercaptoethanol. Prepare fresh and store at 4°C.

5. Nickel-NTA resin elution Buffer B: 0.01 $M$ Tris-HCl, pH 7.4; 0.1 $M$ sodium phosphate; 5 m$M$ β-mercaptoethanol; and 0.5 $M$ NaCl, 20% glycerol. Prepare fresh and store at 4°C.

6. Nickel-NTA resin elution Buffer C: 0.01 $M$ Tris-HCl, pH 6.0; 0.1 $M$ sodium phosphate; 5 m$M$ β-mercaptoethanol; and 1.5 $M$ NaCl, 20% glycerol. Prepare fresh and store at 4°C.

7. Nickel-NTA resin elution Buffer C50: 0.01 $M$ Tris-HCl, pH 6.0; 0.1 $M$ sodium phosphate; 5 m$M$ β-mercaptoethanol; 0.75 $M$ NaCl; 10% glycerol; and 0.05 $M$ imidazole. Prepare fresh and store at 4°C.

8. Nickel-NTA resin elution Buffer D: 0.01 $M$ Tris-HCl, pH 6.0; 0.1 $M$ sodium phosphate; 5 m$M$ β-mercaptoethanol; 0.75 $M$ NaCl; 10% glycerol; and 0.35 $M$ imidazole. Prepare fresh and store at 4°C.

9. Nickel-NTA resin (Qiagen).

10. Thrombin G25 exchange buffer: 50 m$M$ Tris-HCl, pH 7.5; 150 m$M$ NaCl. Store at 4°C.

11. 100 NIH units of Thrombin.

12. 12. Mono Q buffer A: 10 m$M$ potassium phosphate, pH 8.0.

13. Mono Q buffer B: 10 m$M$ potassium phosphate, pH 8.0; 0.5 $M$ NaCl.

14. Dialysis buffer: 50 m$M$ Tris-HCl, pH 7.5; 150 m$M$ NaCl; 1 m$M$ glucose; 1 m$M$ EDTA; 5 m$M$ dithiothreitol (DTT); and 1 m$M$ β-ketoisocaproate.

***2.2. Branched Chain Aminotransferase Assay***

1. Reaction buffer: 25 m$M$ potassium phosphate buffer, pH 7.8.

2. Radioactive Substrate: 1 m$M$ α-keto[1-$^{14}$C] isovalerate and 12 m$M$ isoleucine.

3. Reducing agent: 2 m$M$ DTT.

4. Stock Cofactor: 20 m$M$ pyridoxal phosphate.

5. Stop solution: 5 $N$ acetic acid.

6. Trap for C: 5 $N$ potassium hydroxide.

7. Chemical decarboxylation of unreacted substrate: 30% hydrogen peroxide.

***2.3. Oxidation and Sulfenic Acid Trapping of hBCATm***

1. Reaction buffer: 50 m$M$ N-(2-hydroxyethyl)piperazine-N'-2-ethanesulfonic acid (HEPES), pH 7.2 with 1 m$M$ ethylene diamine tetraacetic acid (EDTA). Store at 4°C.

2. hBCAT enzyme: 7 nmol of purified wild-type and mutant hBCATm (C315S, C318A, C315/C318A), respectively.

3. Hydrogen peroxide concentrations: 0, 1.0 1.5, and 2 m$M$ hydrogen peroxide. Prepare just before use and keep on ice.

4. Sulfenic acid specific reagent: 100 m$M$ stock solution of dimedone prepared in dimethyl sulfoxide (DMSO). This is diluted to a final concentration of 5 m$M$ dimedone prepared in aqueous buffer for each reaction.

5. Enzymatic removal of unreacted hydrogen peroxide: 1 mg/mL catalase.

**2.4. Electrospray Ionization Mass Spectrometry Analysis of Labeled BCAT Proteins**

1. Apollo ultrafiltration device 30 kDa cutoff (Oribtal Biosciences).

2. Micromass Quattro II triple-quadrupole mass spectrometer fitted with an electrospray source.

3. Samples contain 50% acetonitrile with 1% formic acid.

4. 10 m$M$ ammonium bicarbonate.

**2.5. Labeling of Thiol Groups with DTNB**

1. Reaction buffer: 50 m$M$ HEPES, pH 7.0; 1 m$M$ EDTA. Store at 4°C.

2. Thiol-specific reagent: 10 m$M$ DTNB.

3. Exchange column: PD 10 gel filtration column (Amersham Biosciences).

## 3. Methods

Except for the endoplasmic reticulum, the normal redox state of the cytosol of prokaryotic and eukaryotic cells is predominantly a very reducing environment that favors the reduction of thiol groups. In the cell, glutathione maintains a reducing environment and also keeps proteins in their reduced state. It is therefore important to include a reducing agent in the study of proteins with reactive thiol groups. Also, depending on the protein, an anaerobic environment may be preferable. In these cases, it is advisable to work in an anaerobic chamber *(10)* and/or use Tris(2-carboxyethyl)phosphine (TCEP; ref.*16*), which we discuss in this section.

The BCAT enzymes are unique among the aminotransferase enzymes in that they have a structurally important redox active CXXC motif *(17–19)*, and require a reducing environment to remain active. Therefore, during storage, and when possible during experiments, a reducing agent such as DTT should be present to maintain the enzymes in their reduced state. However, this is often impossible because a number of thiol specific reagents (such as DTNB) are not compatible with DTT (*see* **Note 1**). Some advances have been made in this area, where the reducing agent TCEP can be used in place of DTT.

The reagent TCEP is a thiol-free reducing reagent which, like DTT, can reduce disulfide bonds but, unlike DTT, it does not react with certain sulfydryl labeling experiments, e.g., using *N*-ethyl maleimide (NEM). In addition, it is more stable than thiol-containing reducing agents because it is resistant to air oxidation. It is also available as an immobilized TCEP disulfide reducing gel (Pierce Chemical Co.). This offers the advantage of maintaining the sample in a reduced state, without using dialysis or gel filtration prior to experiments when a reduced protein is required. This is particularly useful when working with cysteine sulfenic acids *(16)*.

## 3.1. Purification of hBCATm and the CXXC Mutants

1. The hBCATm expression vector is overexpressed in BL21(DE3) cells as described in Davoodi *et al. (20)*.

2. For hBCATm, cells are harvested from 4 L of culture and the pelleted bacteria suspended in extraction buffer A, containing 0.1 *M* sodium phosphate, pH 8.0; 0.01 *M* Tris-HCl; and 5 m*M* β-mercaptoethanol. This mixture is sonicated for 1-min intervals with 30-s rest periods for a total of 10-min intervals at 70% duty cycle using a Branson model 250 sonifer.

3. The extract is centrifuged for 10 min at 7,800*g* at 4°C. The supernatant is transferred to a conical flask and the equivalent of 4 *M* urea is added (*see* **Note 2**).

4. Resuspend the pellet in extraction buffer AU, 0.01 *M* Tris-HCl, pH 8.0; 0.1 *M* sodium phosphate; 5 m*M* β-mercaptoethanol; and 4 *M* urea and sonicate for 1-min intervals with 30-s rest periods for a total 10-min intervals at 70% duty cycle using a Branson model 250 sonifer.

5. Centrifuge the extract for 10 min at 7,800*g* at 4°C and add this supernatant was to the first supernatant.

6. During the centrifugation steps, Nickel-NTA resin is equilibrated in the extraction buffer, 0.1 *M* sodium phosphate, pH 8.0; 0.01 *M* Tris-HCl; 5 m*M* β-mercaptoethanol; and 4 *M* urea. Resuspended Ni-NTA resin (12 mL) is used for every 4 L of bacteria.

7. The combined supernatants are subsequently transferred to the equilibrated Nickel-NTA resin and allowed to stir at room temperature for 1 h at 4°C.

8. The mixture is pumped into an empty Econo column from BioRad, using a peristaltic pump at a flow rate of 1 mL/min. The column is then washed with 50 mL of extraction buffer.

9. Subsequently, the column is washed sequentially with nickel elution buffers B through to C50 (50 mL). Each wash is retained and an aliquot taken to measure BCAT activity (*see* **Note 3**).

10. Finally, hBCATm is eluted using buffer D containing 350 m$M$ imidazole. 11. Subsequently, 1 m$M$ EDTA is added to the sample (*see* **Note 4;** ref.*20*).

11. The histidine-tag was removed from hBCATm by digestion with thrombin (100 NIH units) at 28°C for 1 h.

12. The final purification step of WT or mutant hBCATm proteins is anion-exchange chromatography using the Mono-Q HR 5/5 anion-exchange column (Pharmacia) pre-equilibrated in 10 m$M$ potassium phosphate, pH 8.0 (*see* **Note 5**).

13. The proteins are eluted selectively using a sodium chloride gradient from 0 to 0.5 $M$ in 10 m$M$ potassium phosphate, pH 8.0, over 20 min at a flow rate of 1.0 mL/min.

14. The purified proteins are then dialyzed at 4°C into storage buffer containing 25 m$M$ Tris-HCl at pH 7.5, 150 m$M$ NaCl, 1 m$M$ glucose, 1 m$M$ EDTA, 1 m$M$ α-ketoisocaproate, 5 m$M$ DTT, and 15% glycerol.

15. The WT hBCATc enzyme could be stored at 4°C for 2 d or at –20°C for 1.5–2 mo. The CXXC mutant proteins were stable at –20°C for 6 mo.

16. The concentration of purified protein was estimated with the Schaffner and Weissmann method *(21)* or determined from the absorbance at 280 nm with the extinction coefficient of 67,600 M/cm per monomer *(20)* (*see* **Note 6**).

17. The final yield is approximately 10 mg of purified protein/L of *E. coli.*

### 3.2. Branched Chain Aminotransferase Assay

1. The standard assay for BCAT activity is performed at 37°C by measuring the formation of [1-[14]C]valine from α-keto[1-[14]C]isovalerate *(22)*.

2. A unit of enzyme activity is defined as 1 µmol of valine formed per min under standard conditions at 37°C. All assays are performed in triplicate.

3. The assay vials should be prepared ahead of time. The standard assay solution (0.5 mL) contains 25 m$M$ potassium phosphate buffer at pH 7.8, 5 m$M$ DTT, 5 m$M$ PLP, 1 m$M$ α-keto[1-[14]C]isovalerate, and 12 m$M$ isoleucine.

4. The standard assay mixture is pre-equilibrated at 37°C for 3 min.

5. The transaminase reaction is initiated by adding 10 µL of the appropriate diluted enzyme to the assay vial and the reaction is allowed to proceed for 3 min at 37°C.

6. The reaction is stopped by addition of 500 µL of a 5 $N$ acetic acid solution.

7. Unreacted radioactive substrate is then removed by adding 500 µL of a 30% hydrogen peroxide solution, which results

in chemical decarboxyation of 2-oxo acids. Volatile $^{14}CO_2$ is trapped by 5N KOH suspended in a hanging well, containing fluted filter paper above the vial.

8. An aliquot of reaction mixture (150 µL) is added to a vial containing 3 mL of scintillation fluid and [1-$^{14}$C]valine measured in a scintillation counter.

**3.3. Spectrophoto metric Analysis of Thiol Groups**

1. For titration of the solvent-accessible thiol groups in reduced or oxidized hBCATm proteins, 10 nmol of protein are exchanged into buffer containing 50 m$M$ HEPES, at pH 7.0, and 1 m$M$ EDTA using a PD10 column (Amersham Biosciences) as described by the manufacturer (*see* **Note 7**).

2. Two nanomoles of protein (*see* **Note 8**) are then incubated with a 100-fold excess of DTNB at room temperature for 10 min (*see* **Note 9**).

3. The absorbance change at 412 nm is monitored and the concentration of free thiol groups calculated from the liberated 2-nitro-5-thiobenzoate (thiolate) dianion (TNB) using a molar extinction coefficient of 14,150 $M^{-1}$ cm$^{-1}$ *(23)*.

4. DTNB titration is performed for quantitative measurement of thiol groups in hBCATs.

5. The molar extinction coefficient of 14,150 $M^{-1}$ cm$^{-1}$ for TNB is used for all calculations *(23)*.

6. A 10 m$M$ DTNB stock solution is prepared and reactions performed in 50 m$M$ HEPES, 0.1 m$M$ EDTA, pH 7.0, for 20 min at room temperature. Final DTNB concentrations were always in 100-fold molar excess over the thiol concentration that can be assayed.

**3.4. Mass Spectrometry Analysis of DTNB-Labeled hBCATm (Evidence for Redox Cycling)**

1. The protein samples, labelled with DTNB as described in Section 3.3, are analyzed on a Micromass Quattro II triple quadrupole mass spectrometer fitted with an electrospray source (*see* **Note 10**).

2. Flow injection analysis is used with the carrier solvent consisting of 50:50 acetonitrile/water with 1% formic acid (*see* **Note 11**).

3. The samples are dialyzed in 10 m$M$ ammonium bicarbonate overnight at 4°C (*see* **Note 12**).

4. Approximately 10% acetonitrile is added to the protein samples, and 5–10 µL (1 nmol of hBCATm) of the sample solution are injected.

5. Each sample required 8–20 scans (3-s scans for 1 min), and the data were processed using MassLynx Version 2.0 and the Maximum Entropy software supplied with the program to generate spectra on the absolute molecular weight scale. An example of typical results is shown in **Fig. 10.1**.

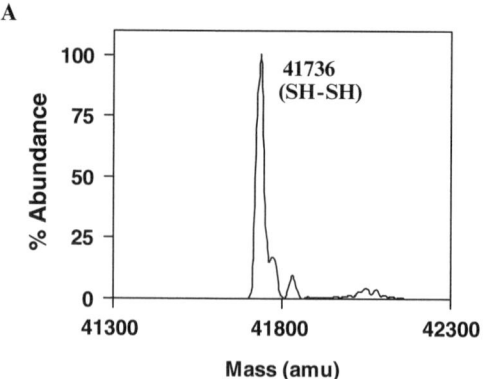

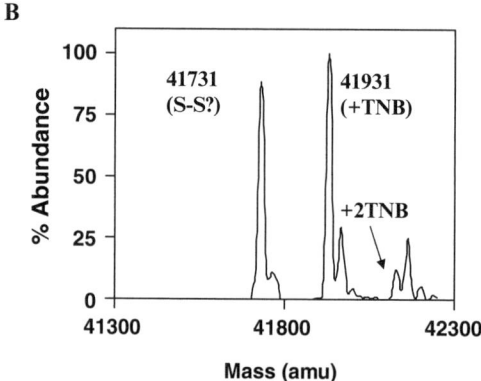

Fig. 10.1. Electrospray mass spectrometry analysis of hBCATm samples in the absence **(A)** or presence **(B)** of DTNB treatment. Human BCATm (control) and the DTNB-labeled protein were dialyzed in 10 m*M* ammonium bicarbonate. The samples were then prepared for mass spectrometry analysis as described in Sect. 3.3. Shown are transformed data, which present the relative abundance of masses for control hBCATm **(A)** and hBCATm incubated with DTNB **(B)**. The major peaks shown in **(A)** (41 736) and **(B)** (41 731 and 41 968) were accompanied by minor peaks at greater molecular weights in each case, which may represent adventitious oxidation with prolonged incubation times. Two minor species in **(B)** likely represent proteins labeled with two TNB moieties (42 124) and a corresponding oxidized species (42 164). Reproduced from **ref**. *(12)*. With permission from American Chemical Society.

### 3.5. Peroxide Sensitivity of hBCATm

1. Human BCATm is exchanged into buffer containing 50 m*M* HEPES, at pH 7.0, and 1 m*M* EDTA.

2. Seven nmol of hBCATm per aliquot are incubated with increasing equivalents of hydrogen peroxide *(1, 2, 4, 8, 16, and 32; see* **Note 13**).

3. The samples are incubated for 24 h at 4°C after which aliquots are taken and the BCAT activity and thiol content measured as described in Sects. 3.2 and 3.3.

4. The samples are incubated for 24 h at 4°C, where the activity was again measured as described in Sect. 3.2.

5. The remainder of the fraction is reacted with a 100-fold molar excess of DTT and allowed to incubate for 18 h at 4°C, after which the activity is again measured (Sect. 3.2).

6. Control samples are incubated under the same sample conditions without the addition of hydrogen peroxide. An example of typical results is shown in **Fig. 10.2.**

7. Confirmation of disulphide bond formation is established with the use of X-ray crystallography *(15,17,18)*. This is illustrated in **Fig. 10.3.**

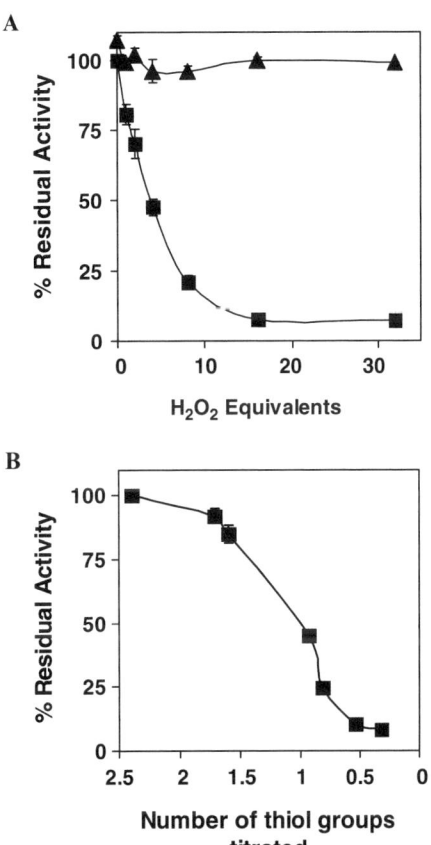

Fig. 10.2. The effect of hydrogen peroxide on the activity and thiol content of hBCATm. The reactions were conducted in 50 m*M* HEPES, pH 7.0, and 1 m*M* EDTA at 4°C for 18–24 h. As shown in **(A)**, residual activity was determined and compared with the number of titratable thiol groups (*n* = 3). In **(B)**, samples were treated with hydrogen peroxide as in panel A (*closed square*) and then incubated with a 100-fold excess of DTT (*closed triangle*) (and reanalyzed for BCAT activity). Reproduced from **ref**. *(12)* with permission from American Chemical Society.

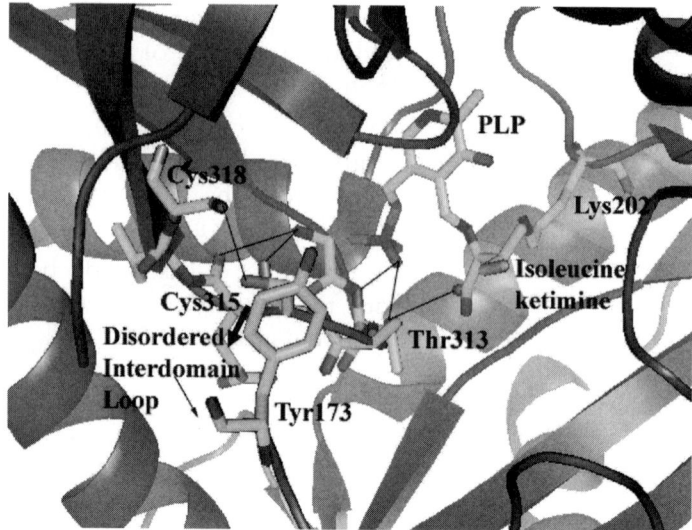

Fig. 10.3. This image represents the active site of monomer B of hBCATm. Evidence of disulfide bond formation between Cys315 and Cys318 is indicated by the arrow. The hydrogen bond interactions are shown by black lines. The thiol group of C315 and C318 are shown in orange and are <3 Å apart, facilitating disulphide bond formation. This figure was generated using Pymol. Reproduced from **ref**. *(12)*. With permission from American Chemical Society.

*3.6. Sulfenic Acid Trapping of hBCATm With Dimedone*

1. Dimedone is not initially water soluble so an organic solvent such as DMSO is used. Before labeling, a stock solution of dimedone (50 m$M$) in DMSO is prepared (*see* **Note 14**).

2. The samples (7 nmol) to be oxidized are initially exchanged into 50 m$M$ HEPES, pH 7.2, and 1 m$M$ EDTA with the use of gel permeation chromatography, with a disposable PD 10 column, according to the manufacturer's instructions (Amersham Biosciences).

3. The concentration of the proteins is determined from the absorbance at 280 nm using the extinction coefficient of 67,600 $M^{-1}cm^{-1}$ per monomer *(20)*.

4. The samples are concentrated using a 7 mL of Apollo ultrafiltration device with a 30-kDa cutoff (Oribtal Biosciences, Topsfield, MA).

5. Dimedone is added to a final concentration of 5 m$M$ to each of the samples containing 4.5 nmol of protein. In addition, a control was included which contained DMSO without dimedone. These samples are brought up to 210 μL with HEPES.

6. Hydrogen peroxide at 0, 1, 1.5, or 2 m$M$ concentrations are subsequently added to the aforementioned samples containing dimedone and incubated at 25°C for up to 23 h.

7. After incubation, catalase (5 µL of a 1 mg/mL stock solution) is added followed by a further 20-min incubation. An aliquot, 15 µL, from each reaction is frozen on dry ice and stored at –80°C until measured for BCAT activity (within 24 h).

8. The remainder of each sample is exchanged into 10 m$M$ ammonium bicarbonate buffer using four rounds of dilution (to 6.5 mL) and re-concentration (to ~50 µL) with the Apollo ultrafiltration devices.

9. Particulate matter is removed by centrifugation and samples are analysed by ESI-MS after addition of acetonitrile to 50% and formic acid to 1%.

10. The mass spectrometry analysis followed the same procedure as described in Sect. 3.4. Examples of these results are shown in **Fig. 10.4**.

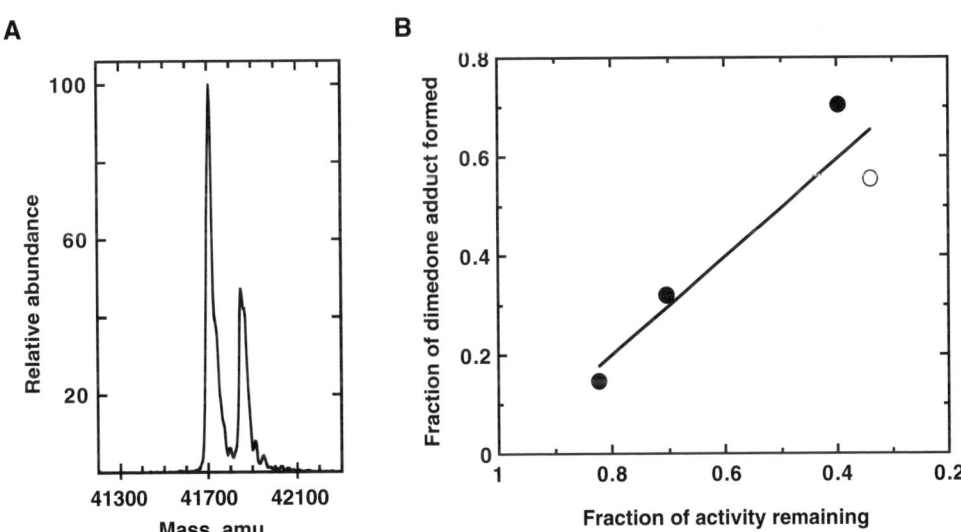

Fig. 10.4. Dimedone adduct formation and activity loss for the H$_2$O$_2$-treated C318A mutant of hBCATm. Samples (4.5 nmol each) were incubated with 1.5 m$M$ (*open circle*) or 2 m$M$ H$_2$O$_2$ (*closed circles*) and 5 m$M$ dimedone at 25°C before the addition of catalase, then assayed for BCAT activity and analysed by electrospray ionization mass spectrometry as described in Section 2. In **(A)**, the transformed mass spectrometry data for the sample after 427 min of treatment are shown, with peaks of maximal abundance at 41,708 and 41,848 atomic mass units for the free and dimedone-adducted protein, respectively. In **(B)**, the correlation between the fraction of dimedone adduct formed and the BCAT activity of the C318A samples from two independent experiments is illustrated, after treatment for 224, 427, 1,180, and 1,375 min (in order of increasing adduct formation; $r$ = 0.936 and slope = –0.987).Reproduced from **ref**. *(12)*. With permission from American Chemical Society.

## 4. Notes

1. A disadvantage with using DTNB as a thiol-specific label is the possibility of generating mixed disulfides, where redox cycling between two structurally accessible cysteine residues can generate two TNB anions as opposed to the expected one TNB anion. This possibility can result in an overestimation of the number of free thiols calculated in a protein. Therefore, it is advisable to react with a second thiol specific reagent independently, which is an irreversible modifier of thiol groups, to ensure that the label is targeting the same thiol groups.

2. Urea is added for this particular protein as it assisted in the binding of the protein to the Nickel column by opening up access to the His-tag on the BCATm protein. This is not always essential, as was seen for purification of BCATc, which did not require Urea in the purification buffers.

3. Aliquots from each fraction should be retained and assayed for activity to ensure maximum binding of your protein to the column.

4. To remove contamination from nickel bleeding from the column 1 mM EDTA is added to the sample. Nickel inhibits BCAT activity and may also influence other metal sensitive proteins.

5. Other buffers compatible with the Mono Q (recommended by the manufacturer) may also be used.

6. Protein determination with the use of ultraviolet absorbance and molar extinction coefficient of 67,600 $M^{-1}\,cm^{-1}$ was used in addition to the amido black assay.

7. A G25 column can be used in place of the PD10 column. EDTA was added to ensure no metal interference to our protein.

8. Two nanomoles of protein was the minimum amount of protein that could be used in order to achieve an accurate estimation of the number of thiol titrated using this method.

9. The thiol-specific reagent DTNB is not particularly soluble in HEPES buffer. Therefore, it was allowed to stir for approx for 1 h (in the dark, because it is light sensitive) and subsequently filtered through a 45-μm filter to remove any insoluble particles. The DTNB solution was then titrated with DTT to calculate the concentration of the solution using the molar extinction coefficient 14,150 $M^{-1}\,cm^{-1}$.

10. Matrix-assisted laser desorption/ionization- or quadrupole-time-of-flight spectrometry may also be used to analyse these samples.

11. The percentage of formic acid can be altered depending on the protein being analysed. However, for BCAT it was observed that percentage less than 1% did not give satisfactory profiles.

12. Dialysis of a sample for mass spectrometry analysis into a buffer such as ammonium bicarbonate is essential as salt containing buffers can interfere with the ionisation of your sample.

13. These doses were used specifically for this protein. Further ranges may be required depending on the system being analysed.

14. NBD-chloride may also be used in place of dimedone to identify sulfenic acids.

## References

1. Droge, W. (2002). Free radicals in the physiological control of cell function. *Physiol Rev 82*, 47–95.

2. Di Simplicio, P., Franconi, F., Frosali, S., and Di Giuseppe, D. (2003). Thiolation and nitrosation of cysteines in biological fluids and cells. *Amino Acids 25*, 323–339.

3. Poole, L. B., Karplus, P. A., and Claiborne, A. (2004). Protein sulfenic acids in redox signaling. *Annu Rev Pharmacol Toxicol 44*, 325–347.

4. Claiborne, A., Yeh, J. I., Mallett, T. C., Luba, J., Crane, E. J., 3rd, Charrier, V., and Parsonage, D. (1999). Protein-sulfenic acids: Diverse roles for an unlikely player in enzyme catalysis and redox regulation. *Biochemistry 38*, 15407–15416.

5. Aslund, F., Zheng, M., Beckwith, J., and Storz, G. (1999). Regulation of the OxyR transcription factor by hydrogen peroxide and the cellular thiol-disulfide status. *Proc Natl Acad Sci USA 96*, 6161–6165.

6. Lee, C., Lee, S. M., Mukhopadhyay, P., Kim, S. J., Lee, S. C., Ahn, W. S., Yu, M. H., Storz, G., and Ryu, S. E. (2004). Redox regulation of OxyR requires specific disulfide bond formation involving a rapid kinetic reaction path. *Nat Struct Mol Biol 11*, 1179–1185.

7. Denu, J. M., and Tanner, K. G. (1998). Specific and reversible inactivation of protein tyrosine phosphatases by hydrogen peroxide: Evidence for a sulfenic acid intermediate and implications for redox regulation. *Biochemistry 37*, 5633–5642.

8. Chiarugi, P., Fiaschi, T., Taddei, M. L., Talini, D., Giannoni, E., Raugei, G., and Ramponi, G. (2001). Two vicinal cysteines confer a peculiar redox regulation to low molecular weight protein tyrosine phosphatase in response to platelet-derived growth factor receptor stimulation. *J Biol Chem 276*, 33478–33487.

9. Dixon, D. P., Fordham-Skelton, A. P., and Edwards, R. (2005). Redox regulation of a soybean tyrosine-specific protein phosphatase. *Biochemistry 44*, 7696–7703.

10. Ellis, H. R., and Poole, L. B. (1997). Novel application of 7-chloro-4-nitrobenzo-2-oxa-1,3-diazole to identify cysteine sulfenic acid in the AhpC component of alkyl hydroperoxide reductase. *Biochemistry 36*, 15013–15018.

11. Poole, L. B. and Ellis, H. R. (2002). Identification of cysteine sulfenic acid in AhpC of alkyl hydroperoxide reductase. *Methods Enzymol 348*, 122–136.

12. Conway, M. E., Poole, L. B., and Hutson, S. M. (2004). Roles for cysteine residues in the regulatory CXXC motif of human mitochondrial branched chain aminotransferase enzyme. *Biochemistry 43*, 7356–7364.

13. Eaton, P., Wright, N., Hearse, D. J., and Shattock, M. J. (2002). Glyceraldehyde phosphate dehydrogenase oxidation during cardiac ischemia and reperfusion. *J Mol Cell Cardiol 34*, 1549–1560.

14. Poole, L. B., Zeng, B. B., Knaggs, S. A., Yakubu, M., and King, S. B. (2005). Synthesis of chemical probes to map sulfenic

acid modifications on proteins. *Bioconjug Chem 16*, 1624–1628.

15. Conway, M. E., Yennawar, N., Wallin, R., Poole, L. B., and Hutson, S. M. (2002). Identification of a peroxide-sensitive redox switch at the CXXC motif in the human mitochondrial branched chain aminotransferase. *Biochemistry 41*, 9070–9078.

16. Baker, L. M., and Poole, L. B. (2003). Catalytic mechanism of thiol peroxidase from *Escherichia coli* sulfenic acid formation and over oxidation of essential CYS61. *J. Biol. Chem. 278*, 9203–9211.

17. Yennawar, N. H., Conway, M. E., Yennawar, H. P., Farber, G. K., and Hutson, S. M. (2002). Crystal structures of human mitochondrial branched chain aminotransferase reaction intermediates: Ketimine and pyridoxamine phosphate forms. *Biochemistry 41*, 11592–11601.

18. Yennawar, N. H., Islam, M. M., Conway, M., Wallin, R., and Hutson, S. M. (2006). Human mitochondrial branched chain aminotransferase isozyme: Structural role of the CXXC center in catalysis. *J. Biol. Chem. 281*, 39660–39671.

19. Conway, M. E., Yennawar, N., Wallin, R., Poole, L. B., and Hutson, S. M. (2003). Human mitochondrial branched chain aminotransferase: Structural basis for substrate specificity and role of redox active cysteines. *Biochim. Biophys. Acta 1647*, 61–65.

20. Davoodi, J., Drown, P. M., Bledsoe, R. K., Wallin, R., Reinhart, G. D., and Hutson, S. M. (1998). Overexpression and characterization of the human mitochondrial and cytosolic branched-chain aminotransferases. *J. Biol. Chem. 273*, 4982–4989.

21. Schaffner, W. and Weissmann, C. (1973). A rapid, sensitive, and specific method for the determination of protein in dilute solution. *Anal. Biochem. 56*, 502–514.

22. Hall, T. R., Wallin, R., Reinhart, G. D., and Hutson, S. M. (1993). Branched chain aminotransferase isoenzymes. Purification and characterization of the rat brain isoenzyme. *J. Biol. Chem. 268*, 3092–3098.

23. Riddles, P. W., Blakeley, R. L., and Zerner, B. (1979). Ellman's reagent: 5,5′-dithiobis (2-nitrobenzoic acid)--a re-examination. *Anal. Biochem. 94*, 75–81.

# Chapter 11

# Detection of Carbonylated Proteins in Two-Dimensional Sodium Dodecyl Sulfate Polyacrylamide Gel Electrophoresis Separations

## Rukhsana Sultana, Shelley F. Newman, Quanzhen Huang, and D. Allan Butterfield

## Abstract

Protein carbonyls are an index of protein oxidation which, in turn, reflects the interplay of oxidative stress and degradation of oxidatively modified proteins. Protein carbonyls are increased in brain proteins in aging and age-related neurodegenerative disorders, including Alzheimer's disease. In this chapter, we outline methods to detect protein carbonyls following two dimensional-based separation of brain proteins.

**Key words:** Imaging analysis, isoelectric focusing, SDS gels, two-dimensional western blots, protein carbonyls.

## 1. Introduction

Oxidative stress has been implicated to play an important role in a number of diseases, including neurodegenerative disorders, cancer, and ischemia. Oxidative stress occurs as the result of an imbalance in the pro-oxidants and antioxidant levels. Reactive oxygen species (ROS), which are produced during oxidative stress, can react with proteins, leading to the formation of protein carbonyls. Protein carbonyls are produced by various reactions, including backbone fragmentation, hydrogen atom abstraction at alpha carbons, attack on several amino acid side-chains (Lys, Arg, Pro, Thr, His, etc.), and by the formation of Michael adducts between His, Lys, and Cys residues and reactive alkenals (e.g., 4-hydroxy-2-nonenal;

John T. Hancock (ed.), *Methods in Molecular Biology, Redox-Mediated Signal Transduction, vol. 476*
© 2008 Humana Press, a part of Springer Science + Business Media, Totowa, NJ
DOI: 10.1007/978-1-59745-129-1_11

ref. *(1))*. Furthermore, the glycation/glycoxidation of Lys amino groups, forming advance glycation end products *(1–5)*, can also lead to the formation of protein carbonyl formation. In addition, a number of reactions of protein radicals can give rise to other radicals, which can cause damage to other biomolecules, e.g., tyrosine-centered free radicals. Protein oxidation could lead to an impairment of wide range of downstream functional consequences, such as dimerization or aggregation, unfolding or conformational changes to expose more hydrophobic residues to an aqueous environment, loss of structural or functional activity, alterations in cellular handling/turnover, effects on gene regulation and expression, modulation of cell signaling, induction of apoptosis and necrosis, and the ubiquitinylation of damaged or aggregated proteins, indicating that protein oxidation has physiological and pathological significance *(1)*.

Protein carbonyls are chemically stable compared with the other products of oxidative stress. Because of this property, protein carbonyls are widely used as markers to assess the extent of oxidation of proteins both in *in vivo* and *in vitro* conditions *(1–3, 5)*. Several sensitive assays were developed for the detection of oxidatively modified proteins *(3, 6–7)*. In our laboratory, we use redox proteomics to identify specific carbonylated proteins in the Alzheimer's disease (AD) and models thereof, in addition to other oxidative stress-related diseases *(8–19)*.

## 2. Materials

### 2.1. Sample Preparation for the Detection of Protein Carbonyls

1. Sample homogenization buffer (pH 7.4): 10 m$M$ $N'$-(2-hydroxyethyl)piperazine-$N'$-2-ethanesulfonic acid (HEPES), 137 m$M$ NaCl, 4.6 m$M$ KCl, 1.1 m$M$ KH$_2$PO$_4$, 0.6 m$M$ MgSO$_4$, 0.5 µg/mL leupeptin (stored as an aliquot at –20°C), 0.7 µg/mL pepstatin (stored as an aliquot at –20°C), 0.5 µg/mL type II S soybean trypsin inhibitor, and 40 µg/mL PMSF in ethanol (prepare fresh).

2. DNPH Solution: 10 m$M$ 2,4-dinitrophenyl hydrazine (Aldrich, Milwakee, WI) prepared in 2 $N$ HCl solutions. This solution can be stored at room temperature.

3. Protein precipitation: Trichloroacetic acetic acid (100%, stored at 4°C).

4. Lipid removal wash: 1:1 (v/v) ethanol/ethyl acetate made fresh before use.

5. BCA reagent kit from Pierce (Rockford, IL).

6. Isoelectric focusing (IEF) tray (BioRad, Hercules, CA).

7. IEF Rehydration Buffer: 8 $M$ urea, 2 $M$ thiourea, 2% CHAPS, 0.2% biolytes, 50 m$M$ dithiothreitol (DTT), and bromophenol blue dissolved in deionized water made fresh before use.

**2.2. IEF of Samples or First Dimension**

1. IEF system (BioRad, Hercules, CA).
2. IPG strips (BioRad).
3. Paper wicks (BioRad).
4. Equilibration trays (BioRad).
5. Unstained and prestained molecular weight marker (Bio-Rad).
6. IEF rehydration buffer: 8 $M$ urea, 2 $M$ thiourea, 2% CHAPS, 0.2% biolytes, 50 m$M$ DTT, and bromophenol blue dissolved in deionized water made fresh before use.

**2.3. Two-Dimensional Electrophoresis**

1. Agarose solution (BioRad).
2. Equilibration trays with lids (BioRad).
3. 2D-gel (BioRad).
4. 1X running buffer (BioRad).
5. Power supply (BioRad).
6. DTT equilibrium buffer (pH 6.8): 50 m$M$ Tris-HCl, 6 $M$ urea, 1% (m/v) sodium dodecyl sulfate (SDS), 30% (v/v) glycerol, and 0.5% DTT dissolved in deionized water made fresh before use.
7. IA equilibrium buffer (pH 6.8): 50 m$M$ Tris-HCl, 6 $M$ urea, 1% (m/v) SDS, 30% (v/v) glycerol, and 4.5% iodoacetamide dissolved in deionized water made fresh before use.
8. Running buffer (10x) is purchased from BioRad and stored at room temperature.
9. Fixing Solution: 10% (v/v) methanol, 7% (v/v) acetic acid dissolved in deionized water and stored at room temperature.
10. SYPRO Ruby Stain: Purchased from BioRad and stored at room temperature.

**2.4. Protein Staining**

1. SYPRO Ruby gel stain (BioRad).

**2.5. Oxyblot (Immunochemical Detection)**

1. Semi-dry transfer unit (BioRad, Hercules, CA).
2. Transfer Buffer: 1% (v/v) 10x running buffer, (no SDS)10% (v/v) methanol diluted with de-ionized water and stored at 4°C.
3. Wash blot/Phosphate Buffered Saline with Tween (PBST): 0.01% (w/v) sodium azide and 0.2% (v/v) Tween 20 dissolved in phosphate buffered saline (PBS) stored at room temperature.
4. Blocking Buffer: 2% bovine serum albumin (BSA) in PBST made fresh before use.

5. Primary Antibody Solution: Anti-dinitrophenyl hydrazone (anti-DNPH) antibody (Chemicon International, Temecula, CA), diluted in PBST (1:100) directly before use.

6. Secondary Antibody Solution: Anti-rabbit conjugated to alkaline phoshatase antibody (Sigma Aldrich, St Louis, MO) diluted in PBST (1:3,000) directly before use.

7. Developing Solution: SigmaFast tablet [5-bromo-4-chloro-3-indolyl phosphate/nitro blue tetrazolium (BCIP/NBT)] (Sigma Aldrich, St. Louis, MO) dissolved in 10 ml DI water, prepared fresh.

8. Nitrocellulose membrane and filter papers (BioRad).

*2.6. Image Analysis*

1. UV transilluminator (Molecular Dynamics, Sunnyvale, CA).

2. Adobe Photoshop on a Microtek Scanmaker 4900.

3. PDQuest image analysis software (BioRad, Hercules, CA).

## 3. Methods

Two-dimensional gel electrophoresis (2D-PAGE) of biological samples involves the separation of proteins based on two physico-chemical properties of proteins, i.e., isoelectric point (PI) and size *(20)*. The first step is IEF, in which the proteins are focused according to their pI. The second step is sodium dodecyl sulfate-polyacrylamide gel electrophoresis (SDS-PAGE), in which the proteins are further separated based on their migration rate. This step provides the two-dimensional map or total protein profile of a given sample. Comparisons of such profiles, or maps, help in determination of differential protein expression. This technique also allows for the screening thousands of proteins at once and thereby providing information about the post-translational modifications, which result in changes in total protein charge (i.e., shift in the position of the protein spot on the gel). In our laboratory, we invoked the use of a parallel analysis in which the 2D gel map is coupled with 2D immunochemical detection of protein carbonyls derivatized by 2,4-dinitrophenylhyrdazine followed by mass spectrometry analysis to identify proteins of particular interest *(8, 11, 12, 15, 16, 18, 21)*. The redox proteomics method used in our laboratory is outlined in **Fig. 11.1.**

*3.1. Sample Derivatization for the Detection of Protein Carbonyls*

1. Prepare 10% (w/v) tissue or cell homogenate in smaple homogenization buffer (10 g of tissue per 100 ml of sample homogenization buffer). Take a small amount of homogenized tissue and sonicate for 10 s on ice.

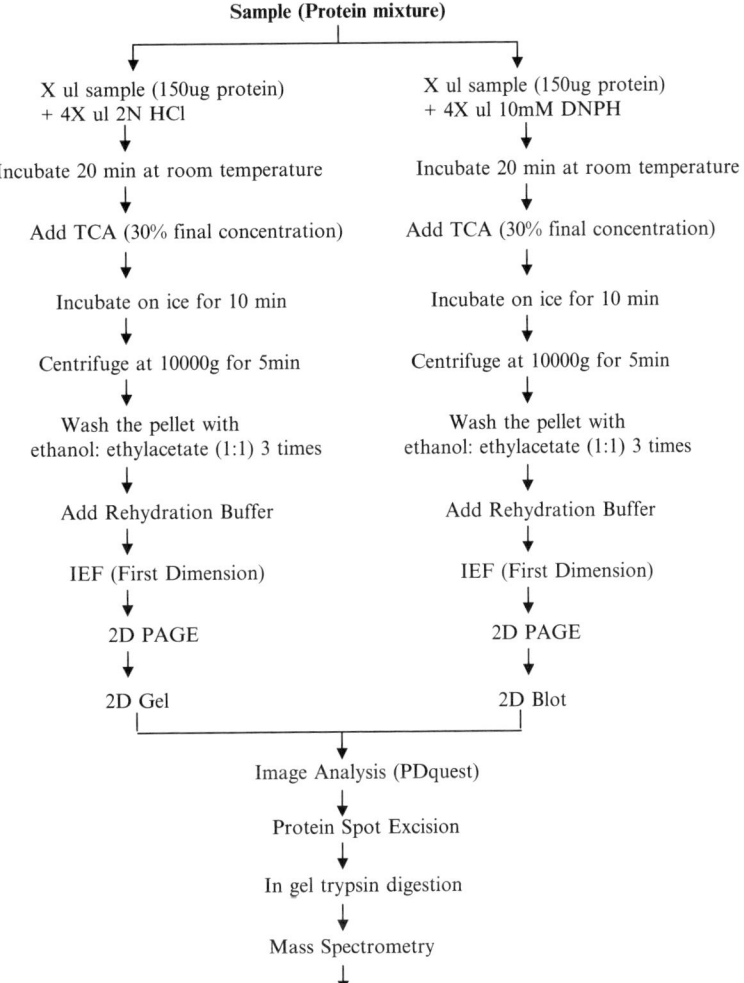

**Sample (Protein mixture)**

| X ul sample (150ug protein) + 4X ul 2N HCl | X ul sample (150ug protein) + 4X ul 10mM DNPH |

Incubate 20 min at room temperature

Add TCA (30% final concentration)

Incubate on ice for 10 min

Centrifuge at 10000g for 5min

Wash the pellet with ethanol: ethylacetate (1:1) 3 times

Add Rehydration Buffer

IEF (First Dimension)

2D PAGE

2D Gel

Incubate 20 min at room temperature

Add TCA (30% final concentration)

Incubate on ice for 10 min

Centrifuge at 10000g for 5min

Wash the pellet with ethanol: ethylacetate (1:1) 3 times

Add Rehydration Buffer

IEF (First Dimension)

2D PAGE

2D Blot

Image Analysis (PDquest)

Protein Spot Excision

In gel trypsin digestion

Mass Spectrometry

Protein database searching for identification of carbonylated proteins

Fig. 11.1. Protocol to identify carbonylated proteins using 2D-gel electrophoresis and western blotting.

2. Determine the protein concentrations in the sonicated samples by using BCA reagent kit.

3. Approx 100–150 μg of the protein is used to derivatize the samples with DNPH for detection of protein carbonyls.

4. To derivatize the samples, add DNPH four times the volume of sample, vortex the samples and incubate at room temperature for 20 min without shaking.

5. Add 100% TCA to the samples to get a final concentration of 30% TCA to precipitate the protein. Incubate the samples on ice for 10 min.

6. Centrifuge the samples at 10,000g for 5 min at 4°C.

7. Decant the supernatant and wash the pellet four times with ice-cold ethanol/ethyl acetate (1:1) mixture. (*see* **Note** 1).

8. All the washing steps should be carried out at 10,000$g$ for 5 min at 4°C.

9. Dry the pellet (*see* **Note** 2).

10. Suspend the pellet in 200 µL of IEF rehydration buffer and incubate at room temperature on a vortex for 1–2 h.

11. Sonicate the samples for 10 s before loading the samples in IEF tray (BioRad).

12. Samples for the protein expression are processed as mentioned previously for the DNPH derivatization except that instead of DNPH, 2 N HCl is added.

### 3.2. IEF of Samples or First Dimension

1. IEF is performed with a BioRad system using 110 mm, pH 3–10-immobilized pH gradient (IPG) strips.

2. Transfer 180 µL of the samples into the bottom of the well in IEF tray carefully by using a micropipet. Avoid bubbles.

3. Pick up the IPG strips with forceps and remove the plastic sheet from the IPG strips. Place IPG strips on top of the sample with gel side facing down. Make sure that the +ve end of strip is towards the +ve end of the IEF tray. This is important for proper operation.

4. Cover the IEF tray and place the IEF tray in the IEF machine and close the lid of the IEF machine. IPG strips are actively rehydrated at 50 V (20°C) overnight.

5. After 1 h open the lid of the IEF machine without turning off the instrument and take the IEF tray outside the machine and add 2 mL of mineral oil in each well to cover the IPG strips (*see* **Note** 3).

6. Place the IEF tray in the IEF machine and close the lid and carryout the active rehydration step for about 16 h.

7. After 16 h, wet the paper wicks with 8 µL of nanopure water and lift the IEF strip using forceps and place a paper wick on both the electrodes (*see* **Note** 4).

8. Place the IEF tray into the machine with right electrodes connection and carry out isoelectrofocusing at 20°C as follows: 300 V for 2-h linear gradient, 500 V for 2-h linear gradient, 1,000 V for 2-h linear gradient, 8,000 V for 8-h linear gradient, and 8,000 V for 10-h rapid gradient.

9. After completion of IEF the IPG strips are transferred to equilibration tray and either processed directly for second dimension or stored in –80°C freezer until use (*see* **Note** 5).

***3.3. Two-Dimensional Electrophoresis (Separation Based in Migration Rate)***

1. Heat agarose solution (*see* **Note** 6)

2. Take the IPG strip out from the –80°C freezer and thaw the IPG strip at room temperature (IPG strips color change from milky white to clear)

3. Incubate the IPG strips in 4 ml of equilibration buffer containing DTT in a disposable equilibration tray with lid with the gels side facing up for 10 min. Keep the equilibration tray in dark (*see* **Note** 7)

4. While waiting for equilibration, prepare 2D-gel by rinsing the gels with DI water (inside and outside). And also prepare 1X running buffer by diluting 100 mL of 10X buffer with 900 mL of deionized water in a measuring cylinder

5. Remove white plastic tape from the bottom of the 2D gels

6. Remove extra water with Kimwipes (tissue paper)

7. After 10 min, transfer the IPG strips into next well in the equilibration tray and add 4 mL of equilibration buffer containing IA, again with the gel side facing up for another 10 min. Keep the equilibration tray in dark (*see* **Note** 8)

8. Dip the IPG strips into 1X running buffer to remove excess equilibration buffer

9. Place the IPG strips with gel side facing up into criterion gels. Do not push the IPG strip into the well

10. Load unstained molecular weight marker into standard well adjacent to the IPG strip well to stain the gels and load precision stained molecular weight markers on the gel that will be used for oxyblot

11. Add warm agarose solution into the wells of criterion gels, avoiding bubbles, and then slowly push the IPG strip on either end till a contact is established between the gel and IPG strip (Strip must be in parallel contact with the gel)

12. Wait for 10 min for agarose to solidify and then place the gels in tank filled with running buffer and then fill the upper tank with running buffer

13. Connect power supply with right connection (+ve to +ve and –ve to –ve)

14. Run the gels at 200 V for 65 min at room temperature, until the dye front (bromophenol blue) exits the gel into the lower tank

15. Disconnect the power supply and disassemble the 2D-apparatus. The gels plates are broken open to remove the gels.

### 3.4. Protein Staining

1. The gels containing non-derivatized proteins with unstained marker is fixed in 50 mL of fixative solution at room temperature for 60 min with gentle agitation.

2. Remove fixative solution and add 50 mL of SYPRO Ruby gel stain and incubate overnight at room temperature on a rocking platform.

### 3.5. Oxyblot (Immunochemical Detection)

1. Gels containing DNPH derivatized proteins are transferred to nitrocellulose membranes for immunochemical detection of protein carbonyls

2. The samples that have been separated by SDS-PAGE are transferred to nitrocellulose membranes using a semidry transfer unit. The gels, nitrocellulose membrane and filter papers are soaked in cold transfer buffer for 10 min.

3. A setup of transfer is prepared in the following order: first place one soaked filter paper on the transfer unit platform, followed by nitrocellulose membrane, gel, and one more filter paper. Roll a glass rod after the gel is placed onto the nitrocellulose membrane to remove bubbles that are trapped between the nitrocellulose membrane, gels, and filter paper. Once the sandwich is ready roll the glass rod once again to ensure no bubbles are is trapped in the transfer sandwich.

4. Connect power supply with right connection (+ve to +ve and –ve to –ve). It is critically important to ensure correct orientation of the power supply or the proteins will move from gel into the filter paper instead of the nitrocellulose membrane.

5. Close the lid of the transfer unit and activate power supply. Transfers are performed at 15 V for 2 h at room temperature. The semitransfer method of proteins not only saves time, but it also requires less amount of transfer buffer.

6. Once the transfer is completed, the transfer unit is disconnected and the unit is carefully disassembled, and the nitrocellulose membrane is taken out. Since pre-stained molecular weight markers are used, we do not need to cut the ends of the nitrocellulose membrane for orientation. The gel and filter papers can then be discarded. The pre-stained molecular weight markers should be clearly visible on the membrane.

7. The nitrocellulose is then incubated in 50 mL of blocking buffer for 1 h at room temperature on a rocking platform. (*see* **Note** 9)

8. To the blocking buffer 1:100 of anti-DNPH antibody is added and incubated for 1 h at room temperature on a rocking platform.

9. The primary antibody is then removed and the membrane washed three times for 5 min each with 50-mL wash blot.

10. The secondary antibody (Anti-Rabbit ALP-conjugated) is freshly prepared for each experiment as 1:3,000 in wash blot and the membrane is incubated on a rocking platform for 1 h.

11. The secondary antibody is discarded and the membrane is washed three times for 5 min each with wash blot.

12. After final wash the blot is developed using Sigma Fast tablet (*see* **Note** 10).

13. Normally, 10–30 min are necessary for color development. After color development, the developer is drained and the membrane is washed with tap water and dried between Kimwipes. An example of the results produced is shown in **Fig. 11.2.**

### *3.6. Image Analysis*

1. SYPRO Ruby stained gels are scanned using a UV transilluminator ($\lambda_{ex}$ = 470 nm, $\lambda_{em}$ = 618 nm).

2. The 2D oxyblots can be scanned with Adobe Photoshop on a Microtek Scanmaker 4900.

3. Oxyblots and 2D gel maps are matched with PDQuest image analysis software to determine the levels of specific protein carbonyls. The carbonyl immunoreactivity of the oxyblot was normalized to the actual protein content as measured by the intensity of a protein stain such as SYPRO ruby.

4. The protein spots showing a significant increase in protein carbonyl levels are excised from the gel and in-gel-digested

**(A)**                                      **(B)**

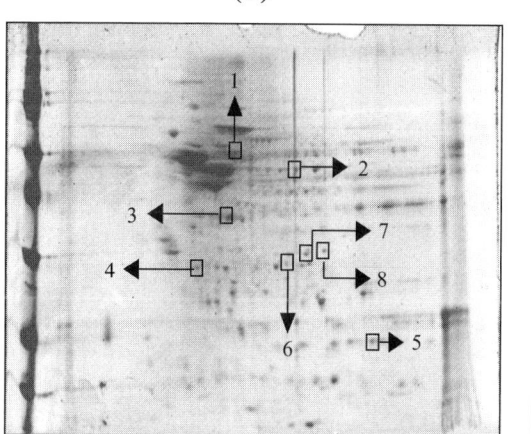
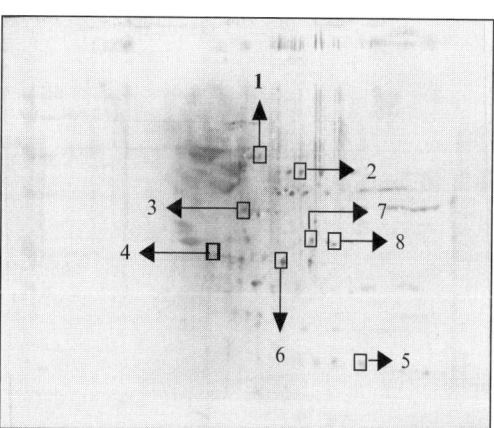

Fig. 11.2. Sypro Ruby-stained gels **(A)** and 2D-Oxyblots **(B)**. Proteins spots showed a significant increase in protein carbonyl immunoreactivity, after normalization of the carbonyl intensities to the protein content are numbered on both gel **(A)** and blot **(B)**. *(1)* Dihydropyrimidinase-related protein 2 (DRP-2); *(2)* alpha enolase; *(3)* gamma-soluble NSF attachment protein; *(4)* ubiquitin carboxy-terminal hydrolase L-1 (UCH L-1); **(5)** Peptidyl-prolyl *cis–trans* isomerase A (Pin 1); *(6)* triose phosphate isomerase (TPI); *(7)* Phosphoglycerate mutase 1 (PGM1); and **(8)** carbonic anhydrase 2 (CA 2).

and submitted for mass spectrometry for mass analysis, finally detected using a peptide search database.

---

## 4. Notes

1. Proper washing of the pellet with lipid removal wash after derivatization is crucial to reduce the background signal on the blot due to excess DNPH. In addition, proper washing removes lipids.

2. Remove the ethanol and ethylacetate solution completely before adding the rehydration buffer, as excess of this mixture could lead to precipitation of proteins.

3. The addition of mineral oil will prevent sample evaporation.

4. Paper wicks will prevent IPG strips from burning.

5. Cover the equilibration tray with polyethylene sheets and wrap the equilibration tray with aluminum foil to prevent moisture formation.

6. Commercially available agarose solution comes with bromophenol blue dye that can be used as a tracking dye to monitor the electrophoresis.

7. DTT can be used to break disulfide bonds.

8. IA can be used to stabilize the disulfide bond once it is cleaved by DTT.

9. Washing is not required between blocking and primary antibody treatment.

10. One tablet of Sigma Fast is dissolved in 10 mL of deionized water.

---

## Acknowledgments

This work was supported in part by grants from NIH [AG-10836;AG-05119].

## References

1. Butterfield, D. A. and Stadtman, E. R. (1997) Protein oxidation processes in aging brain. *Adv. Cell Aging Gerontol.* **2**, 161–191.

2. Berlett, B. S. and Stadtman, E. R. (1997) Protein oxidation in aging, disease, and oxidative stress. *J. Biol. Chem.* **272**, 20313–20316.

3. Dalle-Donne, I., Scaloni, A., and Butterfield, D. A. (2006) *Redox Proteomics: From Protein Modifications to Cellular Dysfunction and Diseases.* Hoboken, NJ: Wiley.

4. Dalle-Donne, I., Scaloni, A., Giustarini, D., Cavarra, E., Tell, G., Lungarella, G., Colombo,

R., Rossi, R., and Milzani, A. (2005) Proteins as biomarkers of oxidative/nitrosative stress in diseases: the contribution of redox proteomics. *Mass. Spectrom. Rev.* **24**, 55–99.

5. Stadtman, E. R. and Levine, R. L. (2003) Free radical-mediated oxidation of free amino acids and amino acid residues in proteins. *Amino Acids* **25**, 207–218.

6. Levine, R. L., Wehr, N., Williams, J. A., Stadtman, E. R., and Shacter, E. (2000) Determination of carbonyl groups in oxidized proteins. *Methods Mol. Biol.* **99**, 15–24.

7. Winterbourn, C. C. and Buss, I. H. (1999) Protein carbonyl measurement by enzyme-linked immunosorbent assay. *Methods Enzymol.* **300**, 106–111.

8. Boyd-Kimball, D., Sultana, R., Poon, H. F., Lynn, B. C., Casamenti, F., Pepeu, G., Klein, J. B. and Butterfield, D. A. (2005) Proteomic identification of proteins specifically oxidized by intracerebral injection of amyloid beta-peptide (1–42) into rat brain: implications for Alzheimer's disease. *Neuroscience* **132**, 313–324.

9. Butterfield, D. A., Perluigi, M., and Sultana, R. (2006a) Oxidative stress in Alzheimer's disease brain: new insights from redox proteomics. *Eur. J. Pharmacol.* **545**, 39–50.

10. Butterfield, D A., Poon, H. F., St Clair, D., Keller, J. N., Pierce, W. M., Klein, J. B., and Markesbery, W. R. (2006b) Redox proteomics identification of oxidatively modified hippocampal proteins in mild cognitive impairment: insights into the development of Alzheimer's disease. *Neurobiol. Dis.* **22**, 223–232.

11. Castegna, A., Aksenov, M., Aksenova, M., Thongboonkerd, V., Klein, J. B., Pierce, W. M., Booze, R., Markesbery, W. R., and Butterfield, D. A. (2002a) Proteomic identification of oxidatively modified proteins in Alzheimer's disease brain. Part I: creatine kinase BB, glutamine synthase, and ubiquitin carboxy-terminal hydrolase L-1. *Free Radic. Biol. Med.* **33**, 562–571.

12. Castegna, A., Aksenov, M., Thongboonkerd, V., Klein, J. B., Pierce, W. M., Booze, R., Markesbery, W. R., and Butterfield, D. A. (2002b) Proteomic identification of oxidatively modified proteins in Alzheimer's disease brain. Part II: dihydropyrimidinase-related protein 2, alpha-enolase and heat shock cognate 71. *J. Neurochem.* **82**, 1524–1532.

13. Castegna, A., Thongboonkerd, V., Klein, J. B., Lynn, B., Markesbery, W. R., and Butterfield, D. A. (2003) Proteomic identification of nitrated proteins in Alzheimer's disease brain. *J. Neurochem.* **85**, 1394–1401.

14. Perluigi, M., Poon, H. F., Maragos, W., Pierce, W. M., Klein, J. B., Calabrese, V., Cini, C., De Marco, C., and Butterfield, D. A. (2005) Proteomic analysis of protein expression and oxidative modification in r6/2 transgenic mice: a model of Huntington disease. *Mol. Cell Proteomics* **4**, 1849–1861.

15. Poon, H. F., Shepherd, H. M., Reed, T. T., Calabrese, V., Stella, A. M., Pennisi, G., Cai, J., Pierce, W. M., Klein, J. B., and Butterfield, D. A. (2006) Proteomics analysis provides insight into caloric restriction mediated oxidation and expression of brain proteins associated with age-related impaired cellular processes: mitochondrial dysfunction, glutamate dysregulation and impaired protein synthesis. *Neurobiol. Aging* **27**, 1020–1034.

16. Poon, H. F., Vaishnav, R. A., Butterfield, D. A., Getchell, M. L., and Getchell, T. V. (2005) Proteomic identification of differentially expressed proteins in the aging murine olfactory system and transcriptional analysis of the associated genes. *J. Neurochem.* **94**, 380–392.

17. Sultana, R., Boyd-Kimball, D., Poon, H. F., Cai, J., Pierce, W. M., Klein, J. B., Markesbery, W. R., Zhou, X. Z., Lu, K. P., and Butterfield, D. A. (2006) Oxidative modification and down-regulation of Pin1 in Alzheimer's disease hippocampus: A redox proteomics analysis. *Neurobiol. Aging* **27**, 918–925.

18. Sultana, R., Boyd-Kimball, D., Poon, H. F., Cai, J., Pierce, W. M., Klein, J. B., Merchant, M., Markesbery, W. R., and Butterfield, D. A. (2006b) Redox proteomics identification of oxidized proteins in Alzheimer's disease hippocampus and cerebellum: an approach to understand pathological and biochemical alterations in AD. *Neurobiol. Aging* **27**, 1564–1576.

19. Sultana, R., Perluigi, M., and Butterfield, D. A. (2006c) Redox proteomics identification of oxidatively modified proteins in Alzheimer's disease brain and in vivo and in vitro models of AD centered around Abeta(1–42). *J. Chromatogr. B. Analyt. Technol. Biomed. Life Sci.* **833**, 3–11.

20. Rabilloud, T. (2002) Two-dimensional gel electrophoresis in proteomics: old, old fashioned, but it still climbs up the mountains. *Proteomics* **2**, 3–10.

21. Sultana, R., Poon, H. F., Cai, J., Pierce, W. M., Merchant, M , Klein, J. B., Markesbery, W. R., and Butterfield, D. A. (2006d) Identification of nitrated proteins in Alzheimer's disease brain using a redox proteomics approach. *Neurobiol. Dis.* **22**, 76–87.

# Chapter 12

# Analysis of Global and Specific Changes in the Disulfide Proteome Using Redox Two-Dimensional Polyacrylamide Gel Electrophoresis

Robert C. Cumming

## Abstract

Protein cysteine sulfhydryl groups are susceptible to a number of redox-dependent modifications, including an interchange between the reduced sulfhydryl and an oxidized disulfide state. A growing body of evidence suggests that reversible disulfide bond formation alters the structure and function of proteins. In this chapter, a method is described for isolating disulfide bonded proteins from different subcellular compartments by using a differential detergent fractionation technique followed by sequential nonreducing/reducing two-dimensional sodium dodecyl sulfate polyacrylamide gel electrophoresis (i.e., Redox 2D-PAGE). This method can be adapted to examine individual redox-active proteins by immunoprecipitating an epitope-tagged redox protein expressed in cultured cells and using Redox 2D-PAGE and mass spectrometry to identify proteins that form mixed disulfides with the tagged protein. With the use of these techniques, it is shown that disulfide bond formation occurs within families of cytoplasmic proteins and may provide a common mechanism used to control multiple physiological processes.

**Key words:** Redox signaling, disulfide bonding, protein folding, two-dimensional electrophoresis, mass spectrometry, epitope-tagging.

## 1. Introduction

Loss of protein cysteine sulfhydryl groups (Cys-SH) can be induced by a wide array of reactive oxygen species (ROS) and is one of the most immediate responses to an elevation in the level of oxidative stress. Functional consequences of Cys-SH loss include protein misfolding, catalytic inactivation, and decreased antioxidant capacity (1, 2). Although protein disulfide bonds were originally thought to form primarily in the oxidizing environment of

John T. Hancock (ed.), *Methods in Molecular Biology, Redox-Mediated Signal Transduction, vol. 476*
© 2008 Humana Press, a part of Springer Science+Business Media, Totowa, NJ
DOI: 10.1007/978-1-59745-129-1_12

the endoplasmic reticulum (ER) in eukaryotic cells, proteins are now known to undergo disulfide bonding in other subcellular compartments, including the cytoplasm and nucleus *(2, 3)*. Most protein cysteine residues have a pKa > 8.0 and, in the reducing environment of the cytoplasm and nucleus, remain protonated at physiological pH. However, a small subset of redox-sensitive proteins possess cysteine residues that exist as thiolate anions at physiological pH as the result of a lowering of their $pK_a$ values by charge interactions with neighboring amino acid residues and are therefore highly vulnerable to oxidation. Some of the products of cysteine oxidation are disulfide bonds (Cys-S-S-Cys), cysteine-sulfenic (Cys-SOH), -sulfinic (Cys-$SO_2H$), and -sulfonic (Cys-$SO_3H$) acids. A number of studies, mainly in yeast and bacteria, have shown that reversible protein disulfide bonding is used as a signaling mechanism in response to oxidative stress *(3)*. However, relatively few studies have identified proteins that undergo reversible disulfide bonding in response to oxidative stimuli in mammalian cells or tissues.

In 1974, Sommer and Traut described a technique, termed diagonal sodium dodecyl sulfate polyacrylamide gel electrophoresis (SDS-PAGE), for analyzing intermolecular disulfide bonds generated between ribosomal proteins in *Escherichia coli* after exposure to methyl-4-mercaptobutyrimidate and hydrogen peroxide ($H_2O_2$) *(4)*. In diagonal SDS-PAGE, a protein mixture is resolved in the first dimension under nonreducing conditions, followed by excision of a narrow gel strip in the sample lane over the entire molecular weight range. This strip is then treated with reducing agents, laid at a right angle across a second gel, and then proteins are resolved under reducing electrophoresis in the second dimension. The resultant resolved proteins that contain no disulfide bonds will lie in a diagonal line on the two-dimensional (2D)-gel, whereas proteins that form disulfide bonds will reside in the off-diagonal zones and appear as distinct spots.

This technique has been used to monitor mixed disulfide bond formation of newly synthesized proteins in the ER *(5, 6)* and to isolate proteins targeted by thioredoxin in plants and cyanobacteria *(7, 8)*. More recently, a variation of the diagonal SDS-PAGE technique (Redox 2D-PAGE) has been used to isolate disulfide-bonded proteins (DSBP) in both oxidant-treated rodent nerve cell cultures *(9)* and cardiac myocytes *(10)*. Remarkably, these two independent studies in which the authors used Redox 2D-PAGE to examine DSBP in the mammalian cytosol of different cell types resulted in the detection of many of the same types of proteins, including those involved in molecular chaperoning, translation, glycolysis, cytoskeletal structure, cell growth, and signal transduction *(9, 10)*. These findings indicate that disulfide bond formation within families of cytoplasmic proteins is a fundamental biological mechanism that may affect multiple physiological processes. The Redox 2D-PAGE method described here should offer an initial basis for exploration of the disulfide proteome.

## 2. Materials

### 2.1. Cell Culture and Detergent Extraction Buffers

1. Dulbecco's Modified Eagle's Medium (DMEM; Invitrogen/Gibco, Grand Island, NY) supplemented with 10% fetal bovine serum (FBS, HyClone, Logan, UT) and 1/200 dilution (v/v) of penicillin/streptomycin stock solution (10,000 units/mL;10,000 mg/ml) (Invitrogen/Gibco, Grand Island, NY).

2. Solution of trypsin (0.05%) and ethylenediamine tetraacetic acid (EDTA; 1 m$M$) from Invitrogen/Gibco.

3. Opti-MEM serum free transfection media and Lipofectamine 2000 reagent (Invitrogen/Gibco).

4. Piperazine-$N,N'$-bis(2-ethanesulfonic acid) (PIPES) 4X stock buffer: 10 m$M$ PIPES, 300 m$M$ sucrose, 100 m$M$ NaCl, 3 m$M$ MgCl$_2$ (*see* **Note** 1).

5. 100 m$M$ EDTA stock solution: Dissolve 3.6 g of EDTA in 100 mL of water (final volume). Store at room temperature.

6. 100 m$M$ phenylmethylsulfonyl fluoride (PMSF) stock solution: Dissolve 174 mg of PMSF in 10 mL of isopropanal and vigorously vortex. Store at room temperature in the dark.

7. Modified digitonin extraction buffer for cytosolic fractionation *(11)*. 10 m$M$ PIPES, pH 6.8; 0.015% (w/v) digitonin; 300 m$M$ sucrose; 100 m$M$ NaCl; 3 m$M$ MgCl$_2$; 5 m$M$ EDTA; 1 m$M$ PMSF; and 40 m$M$ iodoacetamide (IA). Add 25 mL of 4X PIPES stock buffer and 60 mL of water in a small flask with a stir bar and dissolve 18.75 digitonin powder by heating. Once dissolved, add 5 mL of EDTA stock buffer and 1 mL of PMSF stock buffer. Cool the solution to 4°C, adjust pH to 6.8, and bring final volume up to 100 mL with water. The extraction buffer should be aliquoted, fresh-frozen, and stored at –80°C until use. (*see* **Note** 2).

8. Modified Triton X-100 extraction buffer for membrane/organelle fractionation *(11)*. 10 m$M$ PIPES, pH 7.4; 0.5% (v/v) Triton X-100; 300 m$M$ sucrose; 100 m$M$ NaCl; 3 m$M$ MgCl$_2$; 5 m$M$ EDTA, 1 m$M$ PMSF; and 40 m$M$ IA. Add 25 mL of 4X PIPES stock buffer, 1 mL of PMSF stock buffer,

3 mL of EDTA stock buffer, and 5 mL of freshly prepared 10% (v/v) Triton X-100. Cool to 4°C, adjust pH to 7.4 and add ultrapure water to a final volume of 100 mL. Aliquot and store at –80°C. (*see* **Note 2**).

9. SDS nuclear extraction buffer: 50 m$M$ Tris-HCl, pH 7.4; 2% (w/v); 1 m$M$ PMSF; 40 m$M$ IA. Solution should be made fresh for each experiment and kept at room temperature until use to prevent precipitation of SDS.

10. Trichloroacetic acid (Sigma-Aldrich, St. Louis, MO) 100% (w/v) for protein precipitation (avoid contact with skin). Dissolve in ultrapure water and store at 4°C.

11. Protein resolubilization buffer: 100 m$M$ Tris-HCl, pH 8.0; 2% SDS; 1 m$M$ PMSF; and 40 m$M$ IA. Make fresh as required.

12. Lowry detergent compatible protein assay (BioRad, Hercules, CA).

*2.2. Redox Two-Dimensional Polyacrylamide Gel Electrophoresis (Redox 2D-PAGE)*

1. Solutions for Tris/glycine SDS PAGE: 30% acrylamide/bis solution (37.5:1 with 2.6% C) (avoid skin contact by wearing gloves when handling) and $N,N,N,N'$-tetramethyl-ethylenediamine (TEMED, BioRad). 1 $M$ Tris-HCl (pH 6.8), 1.5 $M$ Tris-HCl (pH 8.8), and 10% SDS (w/v); store at room temperature. Ammonium persulfate (APS, BioRad): prepare 10% (w/v) fresh in ultrapure water as required.

2. Running buffer (10X): 240 m$M$ Tris-Base, 192 m$M$ glycine, and 1% (w/v) SDS. Dissolve 29 g of Tris-base, 144 g of glycine, and 10 g of SDS in 800 mL of ultrapure water then adjust final volume to 1,000 mL. 1X running buffer should be approx pH 8.3. Store at room temperature.

3. 2X SDS sample buffer: 80 m$M$ Tris-HCl, pH 6.8; 4% SDS; 20% glycerol; and 0.01% (w/v) bromophenol blue. Make a 200-mL volume, filter sterilize, and store at room temperature.

4. Agarose overlay solution. Make 10 mL of 2% (w/v) low-melt agarose (BioRad) in 1X running buffer fresh by microwaving for 30 s and let cool for several minutes before using.

5. Prestained molecular weight markers. Precision Plus Protein All Blue Standards (BioRad).

6. 1 $M$DL-dithiothreitol (DTT) stock solution. Dissolve 1.54 g of DTT in 10 mL of water, aliquot in 1-mL volumes, and store at –20°C.

7. No. 9 single edged industrial razor blades (Fisher).

*2.3. Silver Staining*

1. Mass spectrometry (MS)-compatible silver stain: (a) Fixer: 50% methanol, 5% glacial acetic acid; (b) Sensitizer: 0.02% (w/v) sodium thiosulfate ($Na_2S_2O_3$); (c) Silver Stain: 0.1% (w/v) silver nitrate ($AgNO_3$); (d) Developer: 0.04% (v/v)

formalin (35% formaldehyde), 2% (w/v) sodium carbonate ($Na_2CO_3$); (e) Stop solution: 5% (v/v) glacial acetic acid. All chemicals supplied by Sigma-Aldrich (St. Louis, MO) and are made fresh in ultrapure water.

2. Nalgene Staining box, 22.5 × 22.5 × 5 cm (Sigma-Aldrich).

**2.4. Mammalian Expression Vectors and Immunoprecipitation Buffer**

1. pRevTre retroviral response vector and pRevTet-On retroviral expression vector (Clontech, Palo Alto, CA) (*see* **Note** 3).

2. Doxycyclin (Clontech, Palo Alto, CA) stock solution. Dissolve at 1 mg/mL in water and filter sterilize. Store at 4°C in the dark, and use within 4 wk. Aliquots can be frozen at –20°C for long-term storage. Fresh doxycyclin should be added to the media for each experiment.

3. MNT buffer: 20 m$M$ 2-($N$-morpholino)ethanesulfonic acid (MES); 100 m$M$ NaCl; 30 m$M$ Tris-HCl, pH 7.5; 1 m$M$ EDTA; 1 m$M$ PMSF; 0.5% Triton X-100. Prepare in small volumes and store at 4°C, stable for 1 mo.

4. Agarose coupled M2 anti-FLAG antibody (Sigma-Aldrich) should be stored at –20°C. Prepare by centrifuging required amount of agarose beads at 5,000$g$ for 30 s and resuspend in 1 mL of MNT buffer containing 0.1% SDS (MNT-S). Wash 2X with MNT-S and resuspend as a 50% slurry in MNT buffer, make fresh as required. FLAG peptide stock (400 µg/mL): resuspend 4 mg of lyophilized FLAG peptide in 10 mL of MNT-S buffer. Store aliquots at –20°C and avoid freeze/thawing.

# 3. Methods

**3.1. Subcellular Fractionation**

1. The murine hippocampal cell line HT22 has been used extensively to examine oxidative stress-induced cell death (*12, 13*). HT22 cells are routinely passaged when approaching confluence with trypsin/EDTA and seeded in 100-mm culture dishes at a density of 1 × 10$^5$ cells/100 mm dish (*see* **Note** 4). For oxidant exposure studies, cells should be seeded at a density of 2.5 × 10$^5$ cells per 100-mm dish and the following day exposed to oxidants. To ensure that enough protein is obtained for each subcellular fraction, three 100-mm dishes (approx 1 × 10$^6$ cells) should be used for each experimental treatment and pooled. As a positive control, cells should be treated with either 1 m$M$ diamide (a disulfide-promoting agent) or 5 m$M$ $H_2O_2$ for 5 min (*see* **Note** 5).

2. After oxidant treatment, wash cells with ice-cold phosphate-buffered saline (PBS) and then add 5 mL of PBS containing 40 m$M$ IA for 5 min to prevent post-lysis oxidation of cysteine residues (*see* **Note** 2).

3. Aspirate PBS and add 150 μL of digitonin extraction buffer containing freshly added IA to each dish. Scrape and pool cells then transfer to appropriately labeled tubes on ice.

4. Rock tubes on ice for 10 min and centrifuge extraction mixture at 480$g$ for 10 min at 4°C. Transfer supernatant (approx 500 μl) containing cytosolic proteins to a new tube.

5. Concentrate the cytosolic fractions by adding 55.5 μL of 100% TCA, vortex, and leave on ice for 30 min. Centrifuge (14,000$g$) at 4°C for 20 min and aspirate the supernatant being careful not to leave any residual liquid.

6. Resuspend TCA precipitated pellet in 100 μL of protein resolubilization buffer (*see* **Note** 6).

7. Resuspend digitonin-insoluble pellet from **step 4** in 0.2 mL of ice-cold Triton X-100 extraction buffer with freshly added IA and incubate on ice with gentle shaking for 30 min. Centrifuge at 5,000$g$ for 10 min and transfer the supernatant (membrane/organelle fraction) to a new tube.

8. Resuspend the Triton X-100-insoluble pellet from **step 7** in 100 μL of SDS nuclear extraction buffer and sonicate briefly (approx three 10-s pulses) until solution is no longer viscous.

9. Centrifuge at 14,000$g$ for 20 min at 4°C and transfer the supernatant (nuclear fraction) to a new tube.

10. All subcellular fractions should be kept on ice and used immediately or stored at –80°C until analysis.

11. Protein concentration for all fractions should be determined with the Lowry assay.

12. To all protein fractions, add an equal volume of 2X SDS sample buffer. Do not add any reducing agents and do not boil the samples before loading.

### 3.2. Redox 2D-PAGE

1. Resolve protein fractions by Redox 2D-PAGE using a SE 600 vertical electrophoresis system (Amersham Biosciences, Piscataway, NJ). Plates (18 × 16 cm) should be scrubbed thoroughly with a rinsable detergent, and then sequentially washed with distilled water and 90% ethanol followed by air drying. Prepare one 1.0-mm thick 10% first dimension gel and four 1.5-mm thick 10% second dimension gels (*see* **Note** 7) by mixing 64 mL of water; 52.8 mL of 30% acrylamide mix; 40 mL of 1.5 $M$ Tris-HCl, pH 8.8; 1.6 mL of 10% SDS; 1.6 mL of APS; and 64 μL of TEMED. Pour approximately 20 mL and 30 mL of separating gel solution for the first

dimension gel and each second dimension gels respectively, allowing enough space for a stacking gel, and immediately overlay with isobutanol. Allow gels to polymerize for 1 h at room temperature.

2. Pour off isobutanol and rinse the top of the gel twice with water.

3. Prepare stacking gel by mixing 13.6 mL of water; 3.4 mL of 30% acrylamide mix; 2.5 mL of 1 $M$ Tris-HCl, pH 6.8; 0.2 mL of 10% SDS; 0.2 mL of 10% APS; and 16 µL of TEMED. Quickly pour the stacking gel and insert Teflon gel combs. Use a 15-well, 1-mm thick comb for the first dimension gel and a 12.1-cm wide, 1.5-mm thick preparative comb for each second dimension gel. Allow to polymerize for 1 h at room temperature. Gels can be used immediately or covered with saran wrap, stored at 4°C, and used the next day.

4. Carefully remove the comb from the first dimension gel and rinse wells with water. Assemble the gel in a electrophoresis unit and block the second upper chamber slot by installing the acrylic buffer dam. Prepare 2 L of 1X running buffer and fill upper and lower buffer chambers.

5. Load 120 µg of protein sample (maximum volume 150 µL) per well, no more than four samples total (*see* **Note** 8). Include one well for a prestained molecular weight marker. Run the gel for 6 h at constant current (25 mA) or until the dye front is approximately 2 cm from the bottom of the gel.

6. After electrophoresis, cut off unused lanes of the gel. For the lanes where samples were loaded, leave the top well strips as these are required for orientating the gel lanes for cutting later on.

7. Take the trimmed gel and soak in 2X SDS sample buffer containing freshly added 100 m$M$ DTT. Rock at room temp for 20 min.

8. Wash the gel two times briefly in 1X running buffer

9. Immerse the gel in 2X SDS sample buffer containing 100 m$M$ IA for 15 min, followed by two brief washes in 1X running buffer.

10. Place the gel, with the well strips oriented at the top, on a piece of Whatman paper soaked in running buffer or a strip of rubber, such as the rubber gasket that comes with the SE 600 casting stand, to prevent the gel from slipping. Lay a ruler beside the gel, measure 12 cm from the bottom of the gel upwards and make a straight cut horizontally across the entire gel using a razor blade. Do not separate the upper and lower halves of the gel. Lay a cutting tool (*see* **Fig. 12.1** and **Note** 9) roughly in the middle of the gel lane and parallel

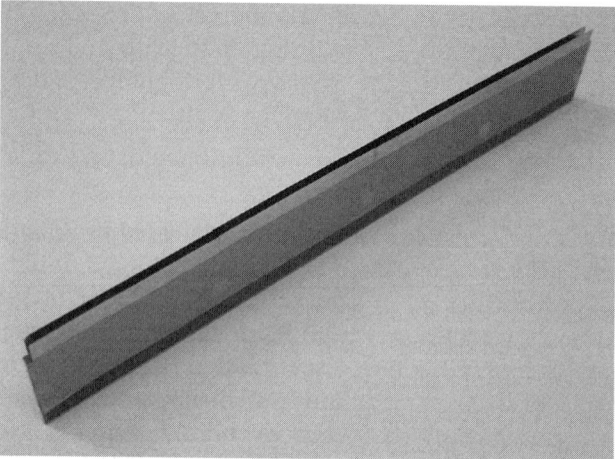

Fig. 12.1. Cutting tool for excising a 1.5-mm wide gel strip. A tool for cutting a narrow strip from the first dimension gel can easily be made by either gluing or taping in series four No. 9 single-edged industrial razor blades to each side of a 16-cm long, 1.5-mm thick spacer. The cutting edges of the blades should protrude approx 3 mm from the edge of the spacer. The cutting tool should be laid on top of the first dimension gel in the middle of the lane and quickly and firmly pushed down on the gel to punch a 1.5-mm wide gel strip for the second dimension gel.

with the well strips. Press quickly and firmly down on the gel and punch out a 1.5-mm gel strip.

11. Remove combs from second dimension gels and rinse wells with water. Leave gels in casting stand as this provides a stable base while inserting gel slices. Add 200 μL of 1X running buffer to cover the top of the preparative well in each second dimension gel.

12. Carefully lift the first dimension gel slice and lay it horizontally in contact with the side of the glass at the top of the second dimension gel. Use a flat end pipette tip to gently push the gel slice into the 1.5-mm thick two-dimensional well, making sure to keep the slice in a horizontal orientation (*see* **Note 10**). Once the slice gets closer to the bottom of the well, use a flat end pipet tip to push the slice in the center until it makes contact with the buffer at bottom of the well, and then move laterally to each side to ensure that no air bubbles get trapped between the slice and the bottom of the well.

13. Make sure that the gel strip is in complete contact with the second dimension gel, then carefully aspirate excess running buffer and using a Pasteur pipette slowly add approx 500 μL of freshly heated and dissolved 2% agarose overlay solution and let harden for 10 min.

14. Assemble gels in electrophoresis unit and add 1X running buffer. Load 3 μl of pre-stained molecular weight standard in marker lane and run at constant current (10 mA/gel)

overnight until the dye front is approximately 2 cm from the bottom of the gel.

**3.3. Silver Stain**

1. Disassemble the gel apparatus and carefully split the plates apart, avoid tearing the gels (*see* **Note** 11). Transfer gels to staining tray (2 gels/tray) containing 500 mL of fixer solution and gently rock for 20 min. Carefully pour off fixer and rinse for 10 min with 50% methanol.

2. Wash gels in water for 2 h, changing water several times.

3. Drain water, add 250 mL of sensitizer solution, and rock for 1 min.

4. Wash gels two times, 1 min each, with water.

5. Stain gels with 250 mL of silver stain solution for 20 min at 4°C with gentle rocking.

6. Wash gels twice, 1 min each wash, with water.

7. Add 250 mL of developer solution and rock gently for 1 min. Quickly drain developer solution and replace with 250 mL of fresh developer solution. Rock for several more minutes until optimal spot intensity/background staining is reached. The diagonal line should appear within 2 min of staining, and off-diagonal (DSBP) spots should appear within 3–5 min after exposure to developer solution. Developing beyond 5 min usually results in high background staining (*see* **Note** 12).

8. Quickly drain developer solution and add 250 mL of stop solution. Rock for 5 min then drain stop solution, leaving several milliliters in staining tray, and add 200 mL of water. Gels should be stored at 4°C and processed for mass spectrometry analysis within 4 days. An example of typically stained gels is shown in **Fig. 12.2**.

9. Spots can be excised, destained, and in-gel digested with trypsin according to standard procedures *(14)*.

**3.4. Immunprecipitation of Eptiope-Tagged Redox-Active Proteins and Isolation of Disulfide-Pairing Partners by Redox 2D-PAGE**

Although the Redox 2D-PAGE method is useful for the profiling of global changes in protein intermolecular disulfide bonding within different subcellular fractions, it does not differentiate between protein complexes linked by either homo- or heterodimeric disulfide bonds. After an initial screen with Redox 2D-PAGE and subsequent MS identification of a particular DSBP of interest, it is possible to determine the nature of the disulfide bonds that the DSBP forms by expressing an epitope tagged version of the protein in cultured cells. We routinely use the FLAG epitope because several antibodies and immuno-affinity matrices are commercially available for detection and purification of FLAG-tagged proteins.

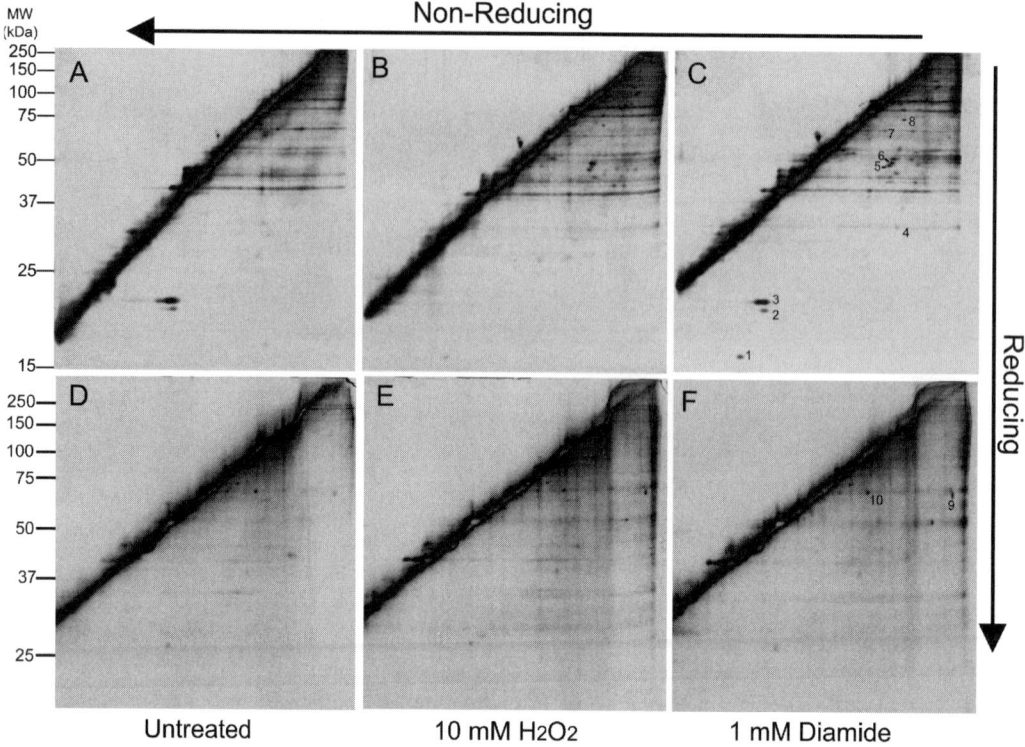

Fig. 12.2. Separation of cytosolic and nuclear DSBP from oxidant-treated HT22 cells by Redox 2D-PAGE. Logarithmically growing HT22 cells were either untreated (**A** and **D**), treated with 10 m*M* H$_2$O$_2$ (**B** and **E**), or 1 m*M* diamide (**C** and **F**) for 5 min. Cytosolic (**A** through **C**) and nuclear (**D** through **F**) protein extracts were prepared under nonreducing conditions and were sequentially resolved by nonreducing and reducing SDS-PAGE followed by silver staining. The prominent diagonal line in each gel represents the majority of proteins that do not form disulfide bonds. Proteins that form intermolecular disulfide bonds exhibit a slower electrophoretic mobility under nonreducing conditions in the first dimension and therefore appear as spots to the right of the diagonal line following reducing SDS-PAGE in the second dimension. A number of spots previously identified by mass spectrometry **(9)** are indicated as reference proteins as follows: 1) nucleoside diphosphate kinase B; 2) peroxiredoxin 2; 3) peroxiredoxin 1; 4) GAPDH; 5) IMPDH-II; 6) histydyl-tRNA synthetase; 7) Hsp 70; 8) cysteinyl-tRNA synthetase; 9) lamin B1; 10) Hsp70. A through C reproduced from **ref. 9.** With permission from the American Society of Biochemistry and Molecular Biology, Inc.

1. Subclone the corresponding cDNA of the DSBP of interest into a suitable mammalian expression vector (*see* **Note 3**).

2. Seed $4 \times 10^5$ HT22 cells per 100-mm dish (*see* **Note 13**).

3. The next day, for each plate, mix 8 µg of plasmid cDNA in 2.5 mL of Opti-MEM media with 10 µL of Lipofectamine 2000 in 2.5 mL of Opti-MEM and leave at room temperature for 20 min.

4. Aspirate media from culture plates and add 5 mL of DNA-Lipofectamine mixture to each plate.

5. Culture transfected cells in an incubator for 6 h, then aspirate DNA/Lipofectamine mixture and replace with

DMEM with 10% fetal bovine serum and place back in incubator.

6. For transient transfections, allow 2 d for cells to recover and express sufficient epitope-tagged protein. For stable Tet-on cell lines, seed cells at $4 \times 10^5$ cells/dish then the next day induce expression with 1 μg/mL doxycyclin for 24 h before harvesting cells.

7. If desired, treat half of the transfected cells with 5 m$M$ $H_2O_2$ for 5 min immediately before harvesting.

8. Wash each plate with PBS then treat with 5 mL of ice-cold PBS + 40 m$M$ IA for 5 min.

9. Aspirate PBS and fractionate cells as described in Sect. 3.1.

10. To precipitated cytosolic protein extracts, resolublize in 100 μL of MNT buffer with 1% SDS and 20 m$M$ IA. Once the protein pellet is fully dissolved, adjust the final volume to 1 mL with MNT buffer (*see* **Note** 14).

11. Add 40 μL of FLAG-agarose (50% slurry) to protein extract and rock overnight at 4°C.

12. The following day, precipitate agarose beads by centrifuging at 5,000$g$ for 30 s. Wash three times with 0.5 mL of MNT-S buffer. Release protein immunocomplexes by competitively eluting with the addition of 100 μL of FLAG peptide (400 μg/mL). Rock at 4°C for 30 min and centrifuge at 10,000$g$ for 30 s. Carefully transfer supernatant to a appropriately labeled tube while avoiding the transference of agarose by using a flat tip pipet. Add an equal volume of 2X SDS sample buffer (with no reducing agents) and use immediately or store at −80°C.

13. Load 150 μL of eluted protein mixture per well in a first dimension gel and resolve and detect immunoprecipitated proteins as described in Sects. 3.2 and 3.3.

14. An example of immunoprecipitated Flag-tagged Hsp70 and its interacting DSBP resolved by Redox 2D-PAGE and silver staining is shown in **Fig. 12.3.**

## 4. Notes

1. For 4X PIPES stock buffer, dissolve 103 g of sucrose, 5.8 g of NaCl and 0.64 g of $MgCl_2 \cdot 6H_2O$ in 200 mL of ultrapure water (double-distilled, deionized, > 18Ω). Solubilize PIPES separately in 12 mL of 1 $N$ NaOH before adding to stock buffer then adjust volume to 250 ml with water. Filter sterilize solution (0.2 μm) and store at 4°C in the dark. Stock solution is stable for up to 2 mo.

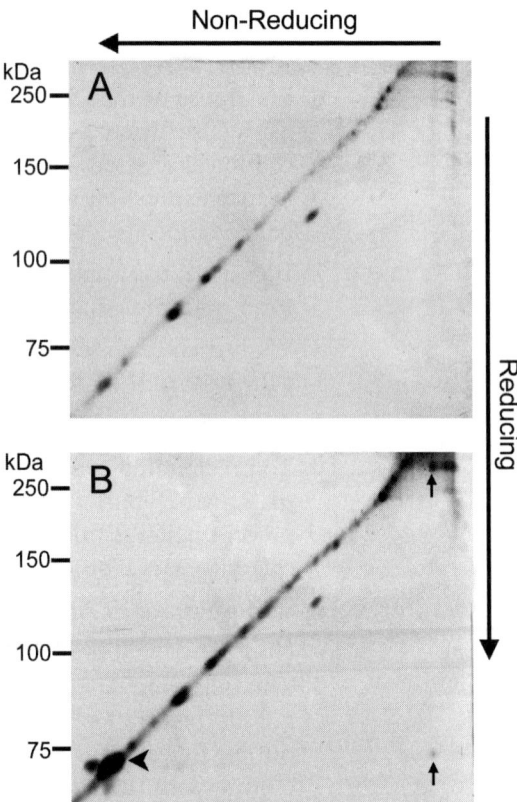

Fig. 12.3. Proteins that form mixed disulfides with epitope-tagged Hsp70 can be isolated with immunoprecipitation and Redox 2D-PAGE. FLAG-tagged Hsp70 **(B)** or the empty vector (pRev) **(A)** was expressed in HT22 cells, and cytosolic extracts were immunoprecipitated under nonreducing conditions with agarose coupled M2 anti-FLAG antibody. Immunoprecipitated complexes were resolved by Redox 2D-PAGE and silver stained. The reduced monomeric form of Hsp70 appears as an intense 70-kDa spot on the diagonal line (indicated by arrowhead). The intermolecular disulfide-linked Hsp70 spot appears to the extreme right of the diagonal line and correlates with a 260-kDa spot directly above it (indicated by arrows). This 260-kDa spot was identified as a mixture of β4-spectrin and APC protein after MS analysis *(9)*. Reproduced from **ref. 9**. With permission from the American Society of Biochemistry and Molecular Biology, *Inc*.

2.  IA is a membrane permeable cysteine alkylating agent that is required to prevent thiol-disulfide exchange and prevent post-lysis oxidation of cysteine residues. Aliquots of both digitonin and Triton-X extraction buffers should be thawed on ice and iodoacetamide (7.4 mg/ml of extraction buffer) added fresh for each experiment.

3.  For stable inducible expression a HT22 cell line is created by transfecting the pRev-Tet-On vector, which allows expression of the reverse tetracycline-controlled transactivator, and a clone is screened for high transactivator expression levels

according to the manufacturers instructions (Clontech, Palo Alto, CA). This stable transactivator line is then transfected with a pRevTRE vector containing a gene of interest under the control of the Tet-response element. Stable clonal cell lines expressing this gene are then selected, following doxycyclin induction, according to the manufactures instructions. Although we have used the tetracycline-inducible system to generate cell lines that stably express an epitope-tagged protein *(9)*, we have also successfully used the mammalian expression vector pcDNA 1.1 (Sigma-Aldrich) for transient expression of epitope-tagged proteins and immunoprecipitation experiments. The FLAG DNA sequence (GATTACAAG-GATGACGACGATAAG) can be combined with a portion of the cDNA sequence of the gene of interest to make a oligonucleotide primer which can then be used in combination with a paired primer to amplify the gene of interest with polymerase chain reaction. Additional restriction sites can be incorporated into the primers and the amplified gene sequence can then be digested with appropriate restriction endonucleases and subcloned into the vector of choice. The resulting vector will then contain an in-frame FLAG sequence. The FLAG tag can be incorporated at either the amino or carboxy terminus of the protein but should be as far from the known or putative reactive cysteine site of the redox-active protein as possible.

4. Care should be taken when culturing HT22 cells as this line can become resistant to oxidative glutamate toxicity if maintained in a confluent state for even a few passages.

5. The choice of oxidant depends on the cell model system and the redox signaling pathway of interest. As a positive control either diamide or $H_2O_2$, at high concentrations, are very effective at inducing disulfide bonding. However, exposure of cells to either agent (1–5 mM) for extended time periods is cytotoxic. Short-term exposure (less than 10 min) should not result in significant cell death.

6. Use a pipet tip to break up pellet. Vortexing and mild heating at 37°C for 5 min may also help resolubilize precipitated protein. Test the pH of a 5-µL aliquot of the protein solution on pH paper. If pH is too low add a small volume (1–2 µL) of 1 *M* Tris-Base, pH 9, to neutralize the residual acid and bring the pH up to approx 7. The protein concentration in both the membrane/organelle and nuclear fractions are generally high enough and do not require TCA precipitation.

7. The first-dimension gel can be poured without using the casting stand by holding the plates and 1-mm spacers together with the clamp assemblies and sealing the bottom of the

gel with 3 *M* yellow electrical tape. Tape must be removed before running the gel. Four second dimension gels can be poured by using an extra set of spacers and a sandwich divider plate between the outer plates when assembled in the casting stand.

8. Frozen protein samples should be thawed at room temperature and vortexed before loading. If a precipitant exists in the protein sample then it can be heated at 40°C for 5 min. The samples should not be boiled or reducing agent added to them.

9. A cutting tool can be made by gluing or taping four No. 9 single edged industrial razor blades on both sides of a 1.5-mm thick spacer (*see* **Fig. 12.1**).

10. Place a few drops of 1X running buffer on top of the gel slice to help it slide down the side of the glass plate. It may be necessary to use the long side of 1 mm spacer to uniformly push the slice into the bottom of the well and prevent the gel slice from flipping on its side.

11. For all steps involving manipulation of gels, it is essential that gloves are worn and frequently changed and that the staining boxes are thoroughly cleaned with rinsable detergent and washed extensively with water to avoid contamination of gels with skin keratin as this will complicate mass spectrometry analysis.

12. The Redox 2D-PAGE technique preferentially results in the detection of abundant proteins in the 40–200 kDa range. Less abundant proteins in the 15–40 kDa range can be detected by developing the silver stained gels for longer periods of time at the expense of a higher background in the upper portion of the gel. Increased resolution of a specific molecular weight range of proteins can be achieved by varying the percentage of acrylamide accordingly in the first and second dimension gels. The number of DSBP is lower in the nucleus compared to the cytoplasm so more protein can be resolved in the first dimension and longer developing times may be required following silver staining for detection of spots (*see* **Fig. 12.2**)

14. To obtain sufficient immunoprecipitated protein complexes for MS analysis, 6–8 dishes (10 cm) per experimental condition are required. As a control, the same number of dishes should be transfected with the empty expression vector.

15. Once the protein pellet is dissolved and volume is brought up to 1 mL, verify that the pH is approximately 7.5 by testing 5-μL aliquots on pH paper. Adjust with 1–5 μL of 1 *M* Tris-Base, pH 9, as required. Determine protein concentration using the Lowry assay and adjust to within 1–2 mg protein/mL. Dilute in MNT-S buffer if necessary.

## References

1. Sitia, R. and Molteni, S. N. (2004) Stress, protein (mis)folding, and signaling: the redox connection. *Sci STKE* **2004**, pe27.

2. Rhee, S. G., Kang, S. W., Jeong, W., Chang, T. S., Yang, K. S., and Woo, H. A. (2005) Intracellular messenger function of hydrogen peroxide and its regulation by peroxiredoxins. *Curr. Opin. Cell Biol.* **17**, 183–189.

3. Linke, K. and Jakob, U. (2003) Not every disulfide lasts forever: disulfide bond formation as a redox switch. *Antioxid. Redox Signal* **5**, 425–434.

4. Sommer, A. and Traut, R. R. (1974) Diagonal polyacrylamide-dodecyl sulfate gel electrophoresis for the identification of ribosomal proteins crosslinked with methyl-4-mercaptobutyrimidate. *Proc. Natl. Acad. Sci. USA* **71**, 3946–3950.

5. Molinari, M. and Helenius, A. (1999) Glycoproteins form mixed disulphides with oxidoreductases during folding in living cells. *Nature* **402**, 90–93.

6. Molinari, M. and Helenius, A. (2000) Chaperone selection during glycoprotein translocation into the endoplasmic reticulum. *Science* **288**, 331–333.

7. Yano, H., Wong, J. H., Lee, Y. M., Cho, M. J., and Buchanan, B. B. (2001) A strategy for the identification of proteins targeted by thioredoxin. *Proc. Natl. Acad. Sci. USA* **98**, 4794–4799.

8. Lindahl, M. and Florencio, F. J. (2003) Thioredoxin-linked processes in cyanobacteria are as numerous as in chloroplasts, but targets are different. *Proc. Natl. Acad. Sci. USA* **100**, 16107–16112.

9. Cumming, R. C., Andon, N. L., Haynes, P. A., Park, M., Fischer, W. H., and Schubert, D. (2004) Protein disulfide bond formation in the cytoplasm during oxidative stress. *J. Biol. Chem.* **279**, 21749–21758.

10. Brennan, J. P., Wait, R., Begum, S., Bell, J. R., Dunn, M. J., and Eaton, P. (2004) Detection and mapping of widespread intermolecular protein disulfide formation during cardiac oxidative stress using proteomics with diagonal electrophoresis. *J. Biol. Chem.* **279**, 41352–41360.

11. Ramsby, M. L. and Makowski, G. S. (1999) Differential detergent fractionation of eukaryotic cells. Analysis by two-dimensional gel electrophoresis. *Methods Mol. Biol.* **112**, 53–66.

12. Tan, S., Sagara, Y., Liu, Y., Maher, P., and Schubert, D. (1998) The regulation of reactive oxygen species production during programmed cell death. *J. Cell Biol.* **141**, 1423–1432.

13. Tan, S., Somia, N., Maher, P., and Schubert, D. (2001) Regulation of antioxidant metabolism by translation initiation factor 2alpha. *J. Cell Biol.* **152**, 997–1006.

14. Shevchenko, A., Wilm, M., Vorm, O., and Mann, M. (1996) Mass spectrometric sequencing of proteins silver-stained polyacrylamide gels. *Anal Chem.* **68**, 850–858.

# Chapter 13

# Protein-Thiol Oxidation, From Single Proteins to Proteome-Wide Analyses

## Natacha Le Moan, Frédérique Tacnet, and Michel B. Toledano

## Abstract

Protein-thiol oxidation subserves multiple biological functions, from enzymatic catalysis to protein oxidative folding, protein trafficking, reactive oxygen (ROS) and nitrogen (RNS) species sensing and signaling and, more generally, protein redox regulation. Protein-thiol oxidation may also constitute a sequel of ROS and RNS toxicity. Accurate and robust methods aimed at monitoring the in vivo redox state of cysteine residues are thus warranted. To this aim, we have developed biochemical approaches that rely on trapping cysteine residues in their in vivo redox state using acidic conditions, followed by the differential labeling of reduced versus oxidized cysteine residues by thiol-specific reagents. These methods have been instrumental in the discovery of eukaryotic peroxide receptors and new ROS-scavenging enzymes and in identifying the repertoire of cytoplasmic oxidized protein thiols. Proteome wide approaches also contributed to establish the functions of the thioredoxin and glutathione pathways in eukaryotic cytoplasmic thiol-redox control.

**Keywords:** Reversible cysteine oxidation, redox regulation, cysteine acid trapping, redox proteomics, ROS sensing and signaling, $H_2O_2$.

## 1. Introduction

The amino acid cysteine has unique chemical properties, endowing it with the ability to engage in a wild variety of redox reactions and to coordinate metals. These properties make cysteine a key residue in enzymatic catalysis, protein oxidative folding and trafficking, reactive oxyten species (ROS) and reactive nitrogen species (RNS) sensing and signaling *(1–3)*. Its unique properties also make this residue vulnerable to the reaction with and modifications by a wide

John T. Hancock (ed.), *Methods in Molecular Biology, Redox-Mediated Signal Transduction, vol. 476*
© 2008 Humana Press, a part of Springer Science+Business Media, Totowa, NJ
DOI: 10.1007/978-1-59745-129-1_13

spectrum of electrophiles, especially ROS and RNS, potentially leading to protein-loss of function.

The cysteine residue exists in vivo in the fully reduced free thiol form (-SH or -S$^-$) and in different oxidation forms, the thiyl radical (-S$^.$), the disulfide bond (Cys-S-S-Cys), the sulfenic (-SOH), sulfinic (SO$_2$H), and sulfonic (-SO$_3$H) acid forms, and the S-nitrosylated form (-S-NO) *(4)*. Cysteine-thiyl radical and cysteine-sulfenic acid are very unstable because of their highly reactive nature and thus cannot be easily identified biochemically. In contrast, cysteine-sulfinic and sulfonic acid are irreversible forms of protein oxidation, although the cysteine-sulfinic acid that forms in peroxiredoxins is enzymatically retro-reduced by sulfiredoxin *(5, 6)*. Disulfide bonds are relatively stable, reversing to the reduced state by thiol-disulfide exchange with kinetics depending on the protein context and the redox nature of the milieu. Disulfide bonds can be formed catalytically by specific thiol oxidases systems, such as the oxygen-dependent FAD-sulfhydryl oxidases Ero1 *(7)* and Erv1 that drive endoplasmic reticulum (ER)-protein oxidative folding during ER secretion and intermembrane mitochondrial space protein import, respectively *(8)*. Disulfide bonds can also form upon reaction of thiols with peroxide or RNS as, for instance, at the reactive cysteine residues of thiol-based peroxidases and of redox sensors *(1, 9, 10)*.

Disulfide bonds are formed as part of the catalytic cycle of specific enzymes such as ribonucleotide reductase *(11)*. In the cytoplasm, disulfide bonds are also often found in the form of mixed disulfides between protein-thiols and GSH also named *S*-thiolation. Protein *S*-thiolation emanates from the condensation between reduced GSH and oxidized protein thiols in the thiyl or sulfenic acid forms that are generated by reaction with peroxides and other intracellular oxidants. *S*-thiolation protects these oxidized cysteine residues from further oxidation and may also regulate the function of specific proteins *(12)*. Two potent NADPH-dependent thiol-reducing systems, the thioredoxin and glutathione (GSH) pathways, are present in the cytoplasm and assist protein-thiol reduction *(13)*. Establishing the in vivo redox state of cysteine residues is thus an important task given the many important cellular responses that rely on cysteine redox modifications.

During cell breakage, reduced cysteine residues can undergo oxidation by O$_2^-$-derived H$_2$O$_2$ and, conversely, oxidized residues can be reduced by thiol-disulfide exchange with cellular reductases, potentially making difficult to evaluate their true in vivo redox state. Acidic quenching of thiol groups, which can circumvent this problem (*see* **Note1**), is best achieved by breaking cells in the presence of trichloroacetic acid (TCA) (pH < 1), which also precipitates soluble cellular proteins. TCA-based acidic quenching is common to and the first step of the methods presented here

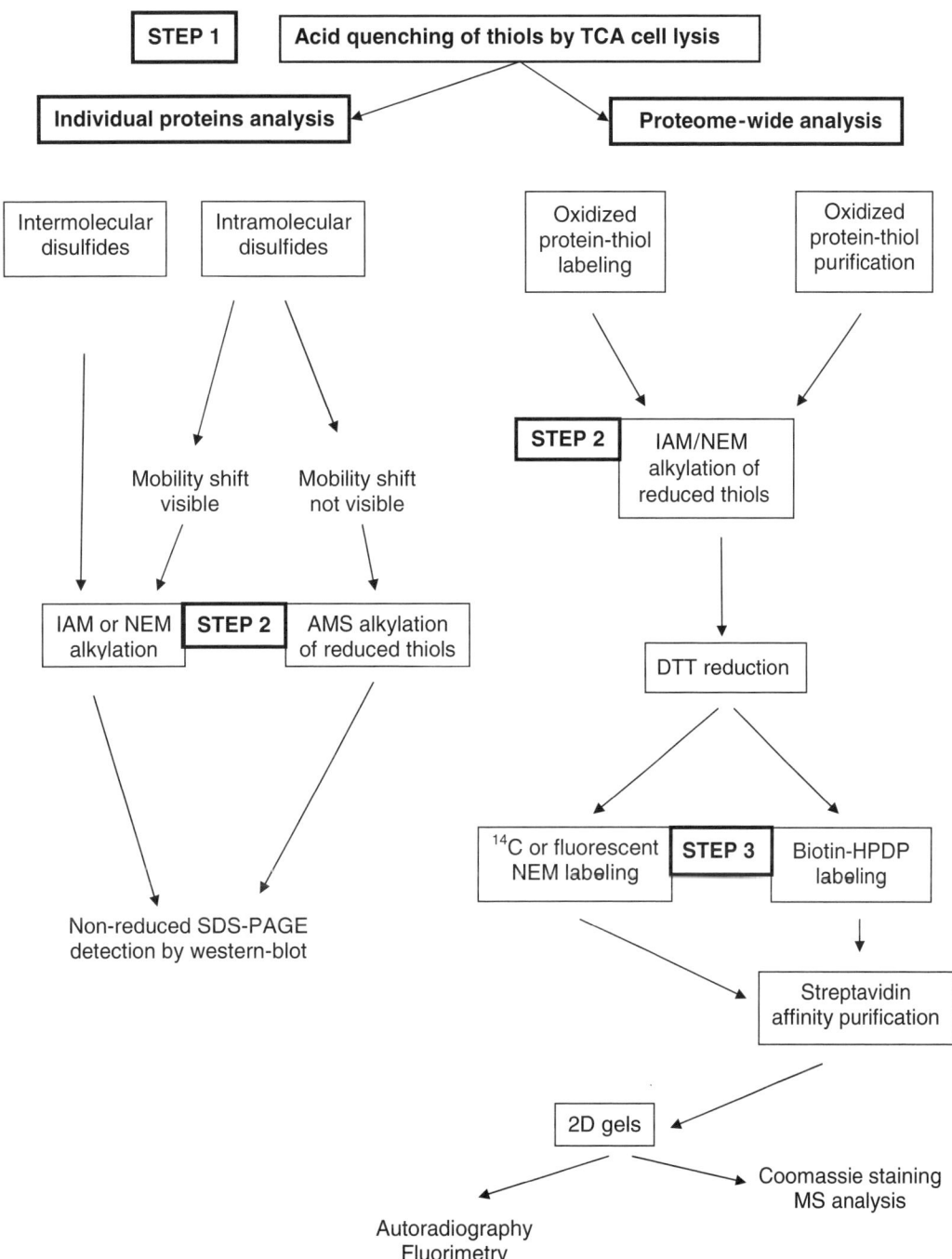

Fig. 13.1. Schematic overview of the different methods presented in this chapter.

(*see* **Fig.13.1**). Basically, the methods presented rely for most of them on the principle of sequential labeling of reduced versus oxidized cysteine residues. Cysteine-specific reagents used for labeling thiols are derived from the alkylating agents iodoacetamide (IAM) or N-ethylmaleimide (NEM) and include the high molecular weight alkylating agents AMS and Mal-PEG, [14]C-radiolabeled-(14, 15) and fluorescent-IAM and -NEM (16, 17). The thiol-reagent HPDP attaching to cysteine residues through a disulfide linkage is also commonly used in fusion with a biotin moiety (18). The redox forms that are accessible to analysis by these techniques are disulfide bonds, whether intra or intermolecular, including S-thiolation and the sulfinic and sulfonic acid forms. Cysteine residues in the sulfenic acid form are very difficult to identify because of their very unstable nature, although some researchers have succeeded in this task, relying either on the exclusive reduction of the sulfenic acid by sodium arsenite (19) or on its reaction with specific chemicals such as dimedone (20).

## 2. Materials

### 2.1. Cell Culture and Cell Lysis

1. CASA medium: 6.7 g/L Yeast Nitrogen Base (Fisher) without amino acids, 20 g/L glucose (Sigma), and 1 g/L casamino acids (Fisher). Adenine, tryptophan, or uracil (5 g/L) are added as required.
2. 500 g of TCA (VWR) is dissolved in 227 mL of water to obtain a 100% solution (stored at 4°C).
3. $H_2O_2$ (Sigma): 30% w/w in $H_2O$ stored at 4°C.
4. Acid-washed glass-beads 425–600 µm in diameter (Sigma).
5. Acetone (VWR) stored at 4°C.
6. Centrifuge (Sigma 3K10).

### 2.2. Sequential Labeling of Oxidized Protein Thiols

1. Denaturing buffer: 25 m$M$ Tris-HCl, pH 8.8, for labeling by [$^{14}$C]NEM and biotin-HPDP, pH 9 or 7.5, for fluorescent-dye labeling; 10 m$M$ EDTA; 8 $M$ urea; and 4% CHAPS. Filter (0.2 µm) and store at –20°C.
2. Iodoacetamide (IAM, Sigma), 50 m$M$ final concentration, is added extemporaneously.
3. Dithiotreitol (DTT, Sigma), 20 m$M$, is added extemporaneously.
4. In the qualitative approach, N-(6-(Biotininamido)Hexyl)-3′-(2′-Pyridylthio) Propionamide (Biotin-HPDP, Pierce) is dissolved at 50 m$M$ in dimethyl sulfoxiode (DMSO) and used at a 0.5 m$M$ final concentration.
5. In the quantitative approach, 1 mCi (27 mM) N-[Ethyl-1-$^{14}$C]-maleimide (NEM) at 37.5 mCi/mmol (Perkin Elmer) is dissolved in 2 mL of DMSO and stored at –80°C until use (2 m$M$ final concentration).

6. For fluorescent labeling, 1 mg of maleimide DYE680 or maleimide DYE780 (Dyomics) is added to 100 μL of DMSO. The stock solution is stored in the dark at –20°C.

7. AMS is 4-acetamido-4′-maleimidylstilbene-2,2′-disulfonic acid, disodium salt.

8. Thermomixer comfort (Eppendorf).

**2.3. Purification of Oxidized Protein Thiols (Biotin-HPDP-Labeled)**

1. Protein concentration is measured using the Micro BCA assay reagent kit (Pierce).

2. Affinity beads: Sepharose CL-4B (Sigma) and Streptavidine-Sepharose High Performance (Amersham Biosciences) are stored at 4°C. Beads are prewashed three times in water and three times in binding buffer, immediately before use.

3. Binding buffer: 25 m$M$ Tris-HCl, pH 8.8; 50 m$M$ NaCl; 5 m$M$ EDTA; 4 $M$ urea, 2% CHAPS.

4. Washing buffer no 1: 25 m$M$ Tris-HCl, pH 8.8;; 1 $M$ NaCl, 8 $M$ urea; 4% CHAPS.

5. Washing buffer no 2: 25 m$M$ Tris-HCl, pH 8.8; 8 $M$ urea; 4% CHAPS.

6. Elution buffer: 7 m$M$ Tris; 8 $M$ urea; 4% CHAPS.

7. Tumbling device.

**2.4. Bi-Dimensional Gel Electrophoresis: First Dimension**

1. Rehydration solution: 8 $M$ urea, 4% CHAPS, 20 m$M$ DTT, 1% immobilized pH gradient buffer, bromophenol blue (0.05% w/v).

2. pH 3–10 nonlinear Immobiline DryStrip gels (18 cm), regular strip holders (18 cm), IPGphor Isoelectrofocusing unit, and IPG buffer 3–10 nonlinear are from Amersham Biosciences.

3. RNAse A (Sigma; 10 mg/ml, i.e., 100x concentrated, stored at –20°C) is added to prevent interference with eventual residual RNAs.

**2.5. Bi-Dimensional Gel Electrophoresis: Second Dimension**

1. Equilibration buffer: 2% SDS, 6 $M$ urea; 50 m$M$ Tris-HCl, pH 8.8; 30% glycerol, bromophenol blue. This buffer is either supplemented with 50 m$M$ DTT or 100 m$M$ IAM, as indicated in the Methods section.

2. Electrophoresis buffer (10X): 250 m$M$ Tris, 1.92 $M$ glycine, 1% SDS.

3. Stabilizing strip buffer: 0.5% (w/v) agarose in electrophoresis buffer.

4. Thirty percent acrylamide/0.4% bis solution (Interchim) (stock solution, stored at 4°C), 1.5 $M$ Tris-HCl, pH 8.8; 10% SDS (0.45-μm filtered).

5. Ammonium persulfate (APS; Sigma), 10% solution in water stored at –20°C and $N,N,N,N'$-tetramethyl-ethylenediamine (TEMED, Sigma).

6. 2-Isopropanol (Fluka) stored at room temperature.

7. Prestained molecular weight marker (Benchmark, Invitrogen).

8. Ettan DALTsix 1 mm Gelcaster, electrophoresis unit and electrophoresis power supply (EPS 601) are from Amersham Biosciences.

9. Laemmli buffer 3X: 150 m$M$ Tris-HCl, pH 6.8; 6% SDS; 30% glycerol; 0.1% bromophenol blue (stored at room temperature).

### 2.6. Detection of Labeled Protein Thiols

1. Speed-Vacuum concentrator.

2. Slab gel dryer (Hoefer).

3. Proteins are fixed in fixation buffer (5:1:5: 5 volumes of water, 5 volumes of ethanol, 1 volume of acetic acid) and stained with Coomassie brilliant blue R250 (Sigma) at 2.5 g/l of fixation buffer. Destain buffer is the same solution without Coomassie blue.

4. [$^{14}$C]NEM-labeled extracts are exposed to general purpose Phosphor screens (Amersham Biosciences), and analyzed using a Phosphorimager with the Image Quant analysis software.

5. Fluorescent extracts are analyzed with an Odyssey infrared imaging system (Li-Cor Biosciences).

## 3. Methods

### 3.1. Methods Outline

An outline of the different methods is presented in **Fig.13.1**. The first step common to all these methods involves breaking cells in the presence of TCA, allowing thiol-acidic quenching, also resulting in the precipitation of soluble cellular proteins. From there, protein-thiol alkylation by IAM, NEM, or their derivatives or with other thiol-specific reagents is performed upon solubilizing TCA-precipitated proteins by increasing the pH of the solution to 7.5–9 and reducing oxidized thiols with DTT (*see* **Note1**). When applicable, differential alkylation of reduced- vs oxidized-cysteine residues is achieved, sequentially before and after reduction of oxidized residues using the thiol-reducing agents DTT or β-mercaptoethanol.

### 3.1.1. Redox State of Individual Proteins In Vivo

The following techniques all rely on the differential separation of oxidized vs reduced proteins by either sodium-dodecyl sulfate polyacrylamide gels electrophoresis (SDS-PAGE) or by isoelectrofocalization and on their visualization by western blot.

They are thus adapted for proteins for which specific antibodies are available. They can be also applied to purified proteins that are then visualized by gel Coomassie staining.

3.1.1.1. Disulfide-Linked
Protein Complexes

In the simplest case, the presence of an intermolecular disulfide bond between the protein of interest and another known or unknown protein can be easily visualized with SDS-PAGE under nonreducing conditions (*see* **Fig. 13.2A,B**). The disulfide-linked complex will migrate at a size corresponding approximately to the sum of the size of the proteins forming this complex. Comparative migration under nonreduced and reduced conditions will establish that the change of mobility of the protein of interest observed under nonreducing conditions is the consequence of a

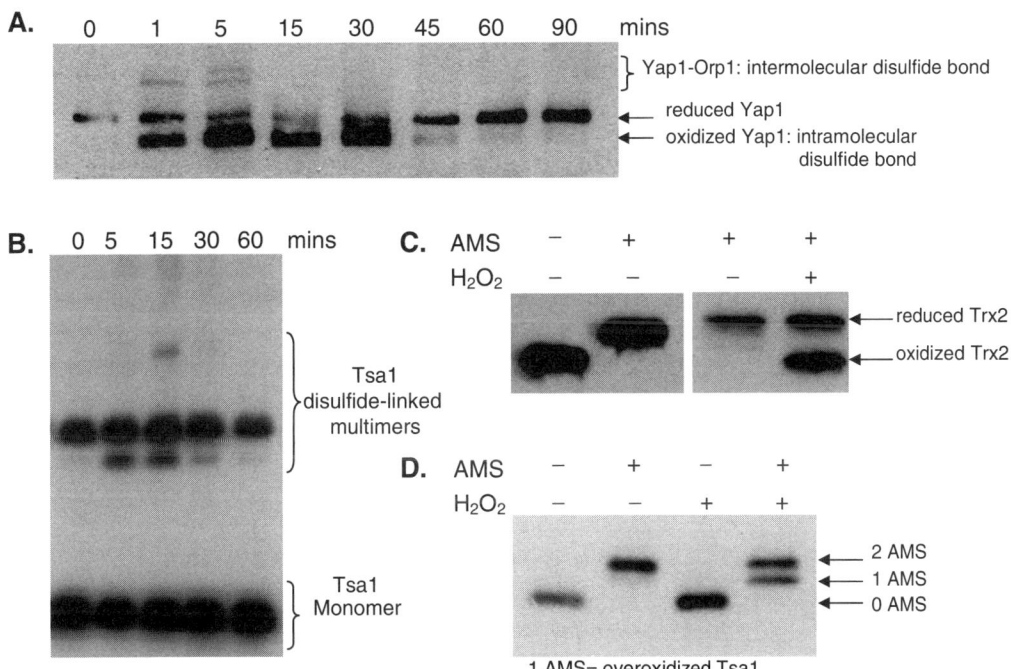

Fig. 13.2. Analysis of the in vivo redox state of individual proteins. **(A)** Yeast extracts of cells expressing a Myc-tagged Yap1 and treated with H₂O₂ (400 µ*M*) during 1–90 min, as indicated, were processed as described in Sect. 3.3.1. Carbamidomethylated Yap1 was analyzed by western blot with the anti-Myc antibody 9E10, revealing the Yap1-Orp1 disulfide-linked complex, the reduced protein and the protein carrying the intramolecular disulfide bonds, as indicated. **(B)** Same as in **(A)** but with cells expressing Myc-tagged Tsa1 (NEM-alkylated). Cells were treated with 500 µ*M* of H₂O₂ during 5–60 min, as indicated. **(C)** Yeast extracts of cells expressing Protein A-tagged Trx2 that were left untreated or were treated with 300 µ*M* of H₂O₂ for 5 min. Samples were processed for AMS alkylation, as described in Sect. 3.3.2. AMS alkylates both Trx2 cysteine residues in reduced protein, but does not modify the intramolecular-disulfide-bonded protein. **(D)** Detection of the overoxidized (cysteine sulfinic acid), DTT-resistant form of Tsa1-Myc. Extracts of cells expressing Myc-tagged Tsa1 are treated or not with 500 µM of H₂O₂ during 30 min, processed as in Sect. 3.3.2 (with a supplementary step of DTT reduction) and separated by reduced SDS-PAGE. Residues alkylated by AMS are those present in vivo in the reduced- and in the disulfide bond DTT-sensitive-forms, but not those in the cysteine-sulfinic acid form. Therefore, both reduced and disulfide-bond forms of Tsa1 will carry two molecules of AMS, whereas the cysteine-sulfinic acid form will carry only one molecule.

reduction-sensitive covalent linkage, which equal to a disulfide bond. Alkylation of remaining reduced cysteine residues with IAM or NEM is absolutely required upon cysteine acid quenching to prevent oxidation of these residues during sample preparation or electrophoresis. This method not only establishes the presence of an intermolecular disulfide linkage but can also accurately quantify the proportion of the protein of interest engaged in this linkage.

### 3.1.1.2. Intramolecular Disulfide Bonds

The presence of an intramolecular disulfide bond may change the protein electrophoretic mobility by maintaining a packed conformation under nonreducing but otherwise-denaturing conditions, especially when the cysteine residues forming the disulfide bond are distant from each other (**Fig.13.2A**). In this case also, protein oxidation will be visualized by the differential mobility of the protein upon reducing vs nonreducing SDS-PAGE. Usually, the intramolecular disulfide bond makes the protein migrates faster *(21)*, but the reverse situation can also be seen *(11)*. Here also, it is essential to covalently block the remaining reduced cysteine residues of the protein by alkylation, as discussed in (1).

### 3.1.1.3. General Method for Analyzing the Redox State of Individual Proteins

More often, however, the change of electrophoretic mobility imparted by an intramolecular disulfide bond is too small to be visualized by SDS-PAGE. In these cases, thiol derivatization is required using modified form of IAM or NEM that increase their molecular weight. A classic reagent that have been used in this laboratory is AMS that consist of a fusion between NEM to another small size molecule, increasing its size by 0.4 kDa, and thus causing a corresponding change of protein SDS-PAGE migration for each cysteine residue it modifies (**Fig.13.2C,D**). MAL-PEG, a fusion between maleimide and methoxypolyethylene glycol, is also often used *(22)*. Many other derivatives of IAM and NEM exist, each having specific advantages. These thiol-labeling reagents can be used to label cysteine residues in their in vivo reduced- or oxidized-forms upon in vitro reduction by DTT.

### 3.1.1.4. Reagents Altering Proteins Isoelectric Points (pI)

IAM and the related compound iodoacetic acid (IAA) differ by their charge. They can be used to differentially label free reduced from oxidized protein thiols. In this case, modification of proteins by either agent is estimated by the differences of charge observed upon isoelectric focalization *(23)*.

### *3.1.2. Proteome-Wide Approaches*

### 3.1.2.1. Oxidized Protein-Thiols Labeling

Oxidized protein thiols can be visualized at the proteome level by differential labeling of reduced vs oxidized cysteine residues upon thiols acid trapping, followed by two-dimensional (2D) gel electrophoresis. Thiol-specific reagents that have been most commonly used are [14]C-labeled IAM and NEM *(14, 15)* (**Fig.13.3**), which allow the identification of oxidized proteins by

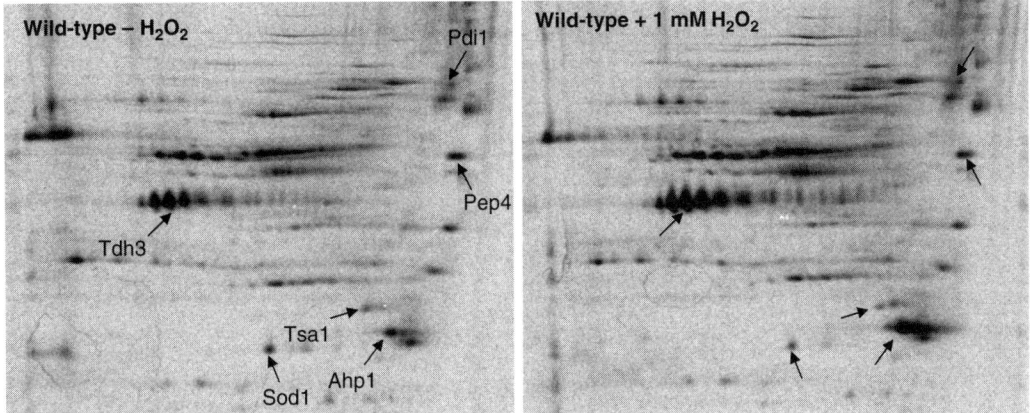

Fig. 13.3. Oxidized protein-thiols were labeled with [14C]NEM and separated by 2D-gel electrophoresis, as described in Sect. 3.2.2. Exponentially growing wild-type yeast cells were either left untreated or were treated during 1 min with 1 m$M$ $H_2O_2$. The glyceraldehyde-3-phosphate dehydrogenase (GAPDH or Tdh3) and the peroxiredoxin Ahp1 increase in oxidation upon $H_2O_2$ treatment.

the radioactive signal of their 2D-gels spot, also providing estimates of protein redox ratios, when accurate controls are used. Fluorescent derivatives of IAM (16), and NEM (17), made by their fusion with fluorescent dyes, require the analysis of 2D-gels by fluorimetry (**Fig. 13.4**). These dyes have the virtue of providing a very accurate measure of the redox ratio of a given substrate, when dyes of different spectral emission are used to differentially label reduced and oxidized cysteine residues respectively. It should be said that these proteome-based thiol-labeling methods are fraught with a relatively high noise background that likely is produced by thiol-nonspecific protein-adduct formation, which warrant the use of the adequate controls to subtract noise from specific signals.

**3.1.2.2. Oxidized Protein-Thiols Purification**

HPDP is an absolute thiol-specific reagent that covalently and reversibly attaches to free thiol groups through a disulfide bridge (14, 18). Labeling of oxidized protein-thiols using biotin-HPDP allows their purification by streptavidin affinity (**Fig. 13.5**). Elution of streptavidin-bound proteins involves reduction of the disulfide linkage between them and biotin-HPDP, using DTT. These purified proteins can then be separated by 2D gel electrophoresis, and identified by mass spectrometry. This method is only qualitative. Biotin-HPDP-modified proteins can also be recognized by streptavidin western blot. Biotin-NEM and Biotin-IAM exist and can also be used for purification of oxidized protein-thiols, or in streptavidin western blots.

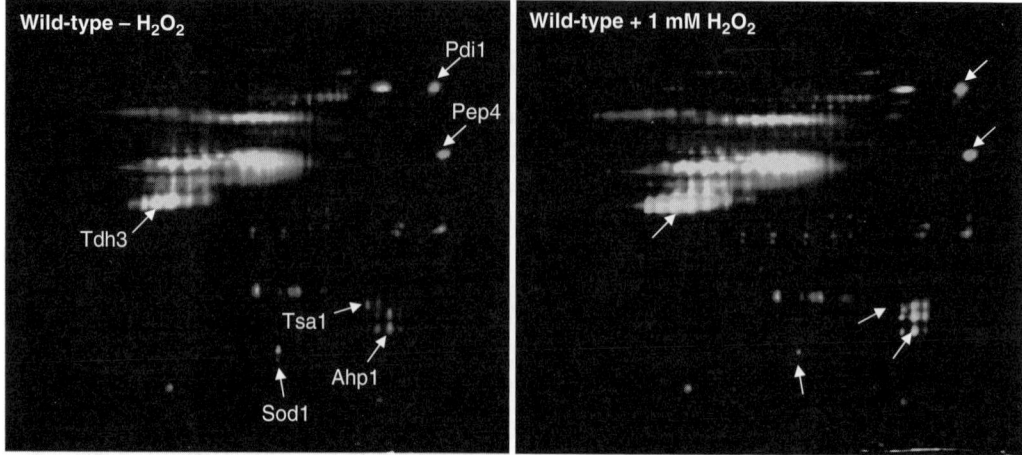

Fig. 13.4. 2D-gel electrophoresis of reduced and oxidized protein-thiols differentially labeled with fluorescent dyes of distinct emission wavelengths, as described in Sect. 3.2.2. Reduced thiols were first labeled with the NEM derivative DYE680 (*red*). Then, after their reduction by DTT, oxidized thiols were labeled with DYE780 (*green*). Red and green signal overlap appears in yellow. The gel on the left corresponds to untreated cells. Pdi1 and Sod1 are here fully oxidized and appear therefore in green. The gel on the right corresponds to cells treated with 1 m*M* of H$_2$O$_2$ during 2 min. The green signals which appeared on the right gel correspond to protein-thiols oxidized by H$_2$O$_2$ (*See* Color Plates).

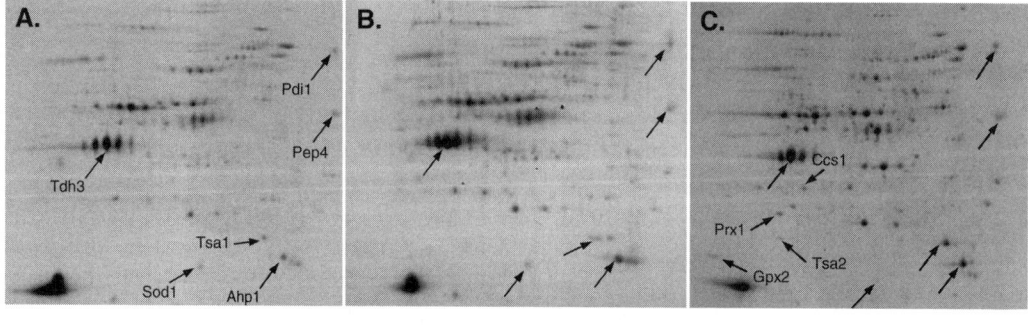

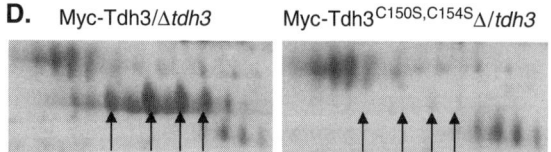

Fig. 13.5. 2D-gel electrophoresis of oxidized protein-thiols purified using biotin-HPDP, as described in Sects. 3.2.2 and 3.2.3. Typical profiles of oxidized protein-thiols purified from wild type cells that were: **(A)** left untreated, or **(B)** treated with 1 m*M* of H$_2$O$_2$ during 1 minute, or **(C)** purified from cells lacking both cytoplasmic thioredoxins. Arrows indicate proteins that show an increased oxidation upon H$_2$O$_2$ treatment or upon inactivation of the thioredoxin pathway. **(D)** A control experiment showing the thiol-specificity of the biotin-HPDP-based purification procedure (From ref. 14): Tdh3 cannot be purified when lacking its two cysteine-residues. A strain with inactivation of the TDH3 gene was transformed with plasmid expressing a Myc-tagged version of Tdh3 (left gel) or Myc-Tagged Tdh3 in which the two cysteines were replaced by serine residues (right gel). The region of gel were Tdh3 migrates is enlarged.

### 3.2. Basic Protocols for Proteome-Wide Oxidized Protein Thiols Identification

The protocol detailed here is adapted for the proteome-wide identification of protein-thiol oxidation in the single-celled eukaryote *Saccharomyces cerevisiae*, using [$^{14}$C]NEM as the thiol-labeling reagent. Modifications required when using other thiol-labeling reagents are indicated with reference to this protocol.

#### 3.2.1. Acid Quenching of Protein Thiol by TCA Precipitation

1. Cells are grown in CASA medium supplemented with adenine, uracil, and tryptophane (5 g/L) at 30°C (200 rpm shaking)
2. TCA is added to a 20% final concentration into a 20-mL cell culture at an O. D. 0.6, which has been treated or not with $H_2O_2$, at concentrations between 0.2 and 1 m$M$ (*see* **Note** 1)
3. The cell culture is centrifuged 5 min at 4,000g and at 4°C
4. The pellet is washed with 1 mL of a 20% TCA cold solution
5. The sample is centrifuged 5 min at 4,000g and at 4°C
6. The pellet is immediately frozen in dry ice
7. 0.2 mL of a 20% TCA cold solution is added to the TCA-precipitated pellet together with 0.1-mL glass beads
8. Cells in suspension in a 1.5-mL Eppendorf are broken by vigorous agitation on a vortex during 10 min. Agitation is repeated three or four times after samples are cooled down by incubation on ice 1 min
9. Glass beads are brought down by gravity during a few minutes. The supernatant is collected in a 1.5-mL Eppendorf tube. The beads are washed with 0.2 mL of 5% TCA cold solution, and the wash solution is added to the first supernatant
10. After centrifugation at 13,000g for 10 min at 4°C, the protein pellet is washed three times in 1 mL of cold acetone (*see* **Note** 2) to remove TCA
11. The pellet is then air-dried in a speed-vacuum for 5 min
12. **Steps 1–11** are identical when using fluorescent NEM or IAM instead of [$^{14}$C]NEM. However, when using biotin-HPDP, the starting culture volume is 500 ml, and **steps 1–11** should be adapted to this volume

#### 3.2.2. Protein-Thiol Labeling (**Figs. 13.3 and 13.4**)

1. Protein pellets are dissolved in 200 μL of denaturing buffer containing 50 m$M$ IAM. Alkylation of free thiols is performed by a 45-min incubation at 30°C under constant shaking (1,400 rpm) (*see* **Note** 3).
2. To remove aggregates, samples are centrifuged 10 min at 13,000g at room temperature.
3. To remove excess of IAM, supernatants are precipitated with cold TCA at a 10% final concentration as in Section 3.2.1, **steps 10** and **11**.

4. Dried pellets are solubilized in 200 μL of denaturing buffer containing 20 mM of DTT. Samples are incubated 45 min at 30°C under shaking to allow full reduction of oxidized thiols.

5. Aggregated proteins are removed by centrifugation 10 min at 13,000g at room temperature.

6. To remove excess of DTT, **steps 3**, **4**, and **5** are repeated once.

7. To label oxidized thiols that have been reduced by DTT in **step 4**, dried pellets are dissolved in 200 μL of denaturing buffer and incubated during 15 min at 30°C, after which 30 μL (15 μCi) of [$^{14}$C]NEM stock solution (2 mM final concentration) is added. Samples are further incubated during 15 min at 30°C under shaking.

8. Protein samples are then centrifuged during 10 min at 13,000g at room temperature to remove aggregated proteins.

9. To remove unbound [$^{14}$C]NEM, repeat **steps 3** and **5**.

10. The dried protein pellets are dissolved in 100 μL of denaturing buffer with shaking at 30°C.

11. After centrifugation, the concentration of the resulting supernatant is determined by the Micro BCA assay reagent kit (*see* **Note4**).

12. 100 μg of the [$^{14}$C]NEM labeled proteins are then subjected to 2D-gel electrophoresis.

13. When using biotin-HPDP, buffer volumes are adapted to the starting culture volume; biotin-HPDP labeling (**step 7**) is performed by incubating samples during 45 min at 30°C, in the dark. Then, **steps 8–10** are identical, except for the volume of denaturing buffer that should be such to optimize protein solubilization (between 500 μl and 3 ml).

14. For fluorescent NEM thiol labeling, dyes are used at a 0.1 mM final concentration in the presence of 200 mM IAM (**step 1** and/or **7**). The pH of the denaturing buffer is 9 in **steps 1** and **4**, and 7.5 in **step 7**. Labeling is performed upon an incubation of 15 min at 4°C in the dark (**step 7**). **Steps 9** and **10** are omitted (*see* **Notes** 5 and 6).

*3.2.3. Purification of Biotin-HPDP Labeled Proteins (Fig. 13.5)*

1. Three to five milligrams of biotin-HPDP-labeled extracts in denaturing buffer from **step 11** (Sect. 3.2.2) are diluted by addition of one volume of buffer containing 50 mM Tris-HCl, pH 8.8; and 100 mM NaCl. Extracts are then precleared by incubation with 500 μL of sepharose CL-4B beads for 3 h at 4°C, with tumbling.

2. The sepharose beads are discarded and 500 μL of streptavidin-sepharose beads are added to the extracts that are then incubated overnight at 4°C with tumbling.

3. The beads are washed once with 1 mL of binding buffer, twice with 1 mL of washing buffer no. 1, and three times with 1 mL of washing buffer no. 2 and once with 1 mL of elution buffer (*see* Sect. 2).

4. The beads are collected. Elution of biotin-HPDP-labeled proteins is achieved in 200 µL of elution buffer containing 20 m$M$ of DTT.

5. The proteins eluted are subjected to 2D-gel electrophoresis.

*3.2.4. 2D-Gel First Dimension*

1. Five micrograms of fluorescent NEM- or 100 µg of [$^{14}$C]NEM-labeled proteins or 200 µL of biotin-HPDP-purified proteins are diluted to a 400 µL final volume of IPG strip rehydration buffer (*see* **Note7**). DTT is omitted from the rehydration buffer in the case of fluorescent NEM-labeled proteins (*see* **Note8**).

2. Protein samples are loaded onto regular strip holders and the Immobiline DryStrip gel pH 3–10 is placed gel-side down against electrodes (*see* **Note9**).

3. Mineral oil is added in the strip holder on top of the strip to prevent sample evaporation and urea crystallization.

4. Sample rehydration is performed by applying 30 V between electrodes on the IPGphor unit during 12 h. Iso-electrofocusing is then performed for 8 h at 20°C with a step-and-hold program using 50 µA per strip (1 h at 150 V, 500 V, 1,000 V for each step and 5 h at 8,000 V).

*3.2.5. 2D-Gel Second Dimension*

1. The polyacrylamide solution (13% acrylamide, 25% Tris-HCl, pH 8.8, 1% APS) is filtered against a 0.45-µ filter before adding 1% SDS and 0.15% Temed (600 mL of solution for six gels).

2. The polyacrylamide solution is poured into the gel caster unit, leaving 0.5 cm space at the top for the IPG strip, and immediately overlaid with 2-propanol.

3. After polymerization, the top of the gel is washed with distilled water. Twenty microliters of the molecular-weight markers solution are applied on a piece of Whatmann of 5 mm$^2$, which is then placed at the gel upper edge on one of its two corners.

4. Isoelectrofocused IPG strips are saturated with SDS for the second dimension: they are first incubated for 15 min with shaking in 10 mL of equilibrating buffer supplemented with 50 m$M$ of DTT and for 15 min in 10 mL of the same buffer supplemented with 100 m$M$ of IAM. For fluorescent NEM-labeled proteins, DTT is omitted.

5. After washing the IPG strip in 1X running buffer, strips are transferred onto the upper edge of polyacrylamide gels and embedded in preboiled stabilizing strip buffer (40°C) (*see* **Note** 10).

6. The gel units are placed in the Ettan Dalt*six* unit containing 4 l of 1X running buffer in the lower chamber and 800 mL of 3X running buffer in the upper chamber.

7. The SDS-PAGE is run first for 30 min at 5 W per gel. Migration is extended either during 18 h at 25°C and at 1.5 W per gel or during 5 h at 15°C and at 17 W per gel.

*3.2.6. Labeled-Proteins Revelation and Analyses*

1. [$^{14}$C]NEM-labeled and biotin-HPDP-eluted proteins 2D gels are fixed during 4 h under constant agitation in 500 mL of fixation buffer, with one wash every hour.

2. [$^{14}$C]NEM labeled proteins-2D gels are washed once in water, placed on Whatmann paper and dried in the gel dryer during 4 h at 80°C. The dried gels are exposed in phosphor-screen cassettes during 3 wk. $^{14}$C-signals are then analyzed on the PhosphorImager using the Image Quant analysis software.

3. Biotin-HPDP-eluted proteins-2D gels are incubated overnight in staining buffer. Oxidized proteins are visualized by destaining gels in destain buffer. Spots of interest are excised and analyzed by mass spectrometry.

4. When using fluorescent NEM, gels are first immersed in water, then carefully placed onto the Odyssey glass screen that had been overlaid with PBS and are scanned.

## 3.3. Protocols for Redox State Analysis of Individual Proteins

*3.3.1. Oxidation Produces a Visible SDS-PAGE Mobility-Shift*
**(Fig. 13.2A,B)**

The following protocol is adapted for proteins carrying intermolecular disulfides bonds, but may also work for those carrying intramolecular disulfide bonds.

1. Initial steps follow **steps 1–11** (Sect. 3.2.1).

2. The speed-vacuum dried pellet is dissolved in 50 μL of [100 m$M$ Tris-HCl, pH 8.8, 1% SDS, 10 m$M$ EDTA] containing 50 m$M$ NEM or 75 m$M$ IAM and incubated 1 h at 37°C or 20 min at 25°C, respectively.

3. Extracts are the centrifuged 10 min at 13,000g.

4. The concentration of proteins in the supernatant is determined by the Micro BCA assay reagent kit. Twenty micrograms of protein extracts are dissolved in 1X Laemmli buffer without reducing agents. This amount of protein extracts has to be adapted for low abundant proteins. Samples are loaded on SDS-PAGE, migrated, and transferred on nitrocellulose membranes for western blot analysis.

*3.3.2. Oxidation Does not Produce a Visible SDS-PAGE Mobility-Shift (v)*

The following protocol is adapted to all oxidized proteins but, in contrast to the previous one, it does not inform on the nature of the oxidation, whether, an inter- or an intramolecular disulfide bond. All steps are performed as in Section 3.3.1, except that in **step 2** NEM is replaced with 15 m$M$ of AMS, and samples are incubated 2 h instead of one.

### 3.4. Discussion

We have presented several variations of the same protocol for the monitoring of the in vivo redox state of proteins. Many other variations are possible depending on the biological questions. These assays have been highly valuable and reliable in our experience. They can serve establishing the oxidized to reduced ratio of a protein, which can then be used to calculate its redox potential, using the Nernst equation. Assays that are geared toward individual proteins not only establish the presence of a redox modification of a given protein, but have also the advantage of providing information on the nature of this modification, especially when combined to site-directed mutagenesis of cysteine residues. For example, these methods have been instrumental in diagnosing two $H_2O_2$-inducible redox forms of the *S. cerevisiae* Yap1 oxidative stress response regulator. These forms are an intermolecular disulfide linkage between Yap1 and a small protein identified as the glutathione peroxidase Orp1 also called Gpx3, which we showed mediate Yap1 oxidation *(24)*, and the intramolecular disulfide bond of the active form of the regulator *(21)* (**Fig.13.2A**). These methods also enabled to discover the function of the sulfinic acid reductase sulfiredoxin *(5)*, by identifying a disulfide-linked complex between this protein and its substrate the peroxiredoxins Tsa1 and Tsa2, upon purification of sulfiredoxin under nonreducing conditions (**Fig.13.2B**). AMS alkylation of reduced thiols has also been instrumental in demonstrating both the in vivo reversibility of the sulfinic acid formed at the peroxiredoxin catalytic cysteine and the catalytic role of sulfiredoxin in reversing this cysteine oxidized form (**Fig.13.2D**).

Proteome-wide approaches can establish the presence of one or more oxidized cysteine residues in a given protein, but not the nature of the oxidation. Despite this limit, the combined use of [$^{14}$C]NEM-labeling (**Fig.13.3**) and biotin-HPDP-based purification of oxidized proteins-thiols (**Fig.13.5**) has enabled us to identify more than 60 oxidized proteins in the cytoplasm of yeast cells *(14)*. This study also established striking differences in the role of the two thiol-redox control pathways glutathione and thioredoxin, with the former carrying the function of cellular thiol-redox buffer and the later having an exclusive role in peroxide catabolism. As an alternative to the proteome-wide approaches presented, the use of diagonal electrophoresis, which consists in consecutive non-reducing and reducing SDS-PAGE, allows assessing the presence of intra- and intermolecular disulfide bonds *(25, 26)*.

Limits to the methods presented are nevertheless present, especially with proteome-wide approaches that restrict analyses to abundant proteins ignoring most regulatory proteins, which are usually expressed at low levels. This question remains unsolved for the moment. Hence, future efforts should be spent at designing new methods allowing the detection of protein of low abundance. The use of two-dimensional difference gel electrophoresis *(27, 28)* through the labeling of thiols using thiol-reagents of different spectral emissions, allows to quantify the differences of protein-oxidation between samples, and should also help increase the coverage of proteomes when combined to mass spectrometry analysis of complex protein solutions. The proteomic analysis of oxidized protein-thiols labeled with either radioactive or fluorescent thiol-reactive reagents is also fraught with a significant level of nonspecific signals, because of the formation of nonspecific amino acid residues-adducts. Efforts should also be made toward new reagents with improved thiol specificity and capable of discriminating between disulfide bonds and irreversible forms of cysteine oxidation such as the sulfinic and sulfonic acid. Because many of the oxidized protein-thiols correspond to protein modification by *S*-thiolation, methods aimed at recognizing the GSH moiety in these complexes are also highly desirable.

## 4. Notes

1. The thiol group can engage in redox reaction only when in the thiolated (deprotonated) state (-S$^-$), which occurs when the pH of the solution > pKa value of the cysteine residue. Free cysteine has a pKa of 8.3, and cysteine residues have pKa values from 4 to 10 depending on their amino acid environment. Therefore cell lysis under acidic conditions, which is best achieved with TCA (pH < 1), will keep cysteine residues in their protonated form thus preventing any redox modifications.

2. During acetone washing, the protein pellet should be manually brought to a powder with the help of a spatula.

3. Denaturing buffer containing urea should not be used at a temperature greater than 30°C, which may change protein pI.

4. A low protein concentration can be the result of partial protein during the acetone wash or to low cells lysis efficiency.

5. Fluorescent dyes are of non-negligible sizes. When using these reagents, protein spot signals very often appear as doublets of

slightly different molecular size (**Fig.13.4**), which can be accounted by an uneven number of modified thiols per protein. The use of dyes at concentrations that saturate protein-thiols should circumvent this problem. However this is not possible due to the price of the reagents. We therefore recommend instead using a higher concentration of IAM.

6. Alkylation temperature at 4°C decreases the nonspecific fluorescent labeling.

7. The presence of bromophenol blue, which migrates towards the anode, allows monitoring the quality of protein sample migration during isoelectrofusing.

8. DTT is omitted during both first and second dimension to prevent subsequent dye labeling of any remaining oxidized thiols by the dyes that are not removed in this particular procedure.

9. Air bubbles under the IPG strip should be avoided, because they perturb Iso-electro-focusing.

10. The plastic side and not the gel side of the IPG strip must be placed against the glass plate. When placing the IPG strip on the upper edge of the SDS polyacrylamide gel, it is also recommended to avoid air bubbles.

## Acknowledgments

Many thanks to Benoit Biteau for the Myc-Tsa1 western-blot, and Ludivine Monceau for the Myc-Yap1 western-blot, and to grass from, ARC and ANR to MBT.

## References

1. Forman, H. J., Fukuto, J. M., and Torres, M. (2004) Redox signaling: thiol chemistry defines which reactive oxygen and nitrogen species can act as second messengers. *Am J Physiol Cell Physiol*, **287**, C246–256.

2. Jacob, C., Giles, G. I., Giles, N. M., and Sies, H. (2003) Sulfur and selenium: the role of oxidation state in protein structure and function. *Angew Chem Int Ed Engl* **42**, 4742–4758.

3. Nathan, C. (2003) Specificity of a third kind: reactive oxygen and nitrogen intermediates in cell signaling. *J Clin Invest* **111**, 769–778.

4. Poole, L. B., Karplus, P. A., and Claiborne, A. (2004) Protein sulfenic acids in redox signaling. *Annu Rev Pharmacol Toxicol* **44**, 325–347.

5. Biteau, B., Labarre, J., and Toledano, M. B. (2003) ATP-dependent reduction of cysteine-sulphinic acid by *S. cerevisiae* sulphiredoxin. *Nature* **425**, 980–984.

6. Woo, H. A., Chae, H. Z., Hwang, S. C., Yang, K. S., Kang, S. W., Kim, K., and Rhee, S. G. (2003) Reversing the inactivation of peroxiredoxins caused by cysteine sulfinic acid formation. *Science* **300**, 653–656.

7. Tu, B. P., and Weissman, J. S. (2002) The FAD- and O(2)-dependent reaction cycle of Ero1-mediated oxidative protein folding in the endoplasmic reticulum. *Mol Cell* **10**, 983–994.

8. Mesecke, N., Terziyska, N., Kozany, C., Baumann, F., Neupert, W., Hell, K., and Herrmann, J. M. (2005) A disulfide relay system in the intermembrane space of

mitochondria that mediates protein import. *Cell*, **121**, 1059–1069.

9. Linke, K., and Jakob, U. (2003) Not every disulfide lasts forever: disulfide bond formation as a redox switch. *Antioxid Redox Signal* **5**, 425–434.

10. Toledano, M. B., Delaunay, A., Monceau, L., and Tacnet, F. (2004) Microbial $H_2O_2$ sensors as archetypical redox signaling modules. *Trends Biochem Sci* **29**, 351–357.

11. Camier, S., Ma, E., Leroy, C., Pruvost, A., Toledano, M. B., and Marsolier-Kergoat, M. C. (2007) Visualization of ribonucleotide reductase catalytic oxidation establishes thioredoxins as its major reductants in yeast. *Free Radic Biol Med*,(in press).

12. Ghezzi, P. (2005) Regulation of protein function by glutathionylation. *Free Radic Res* **39**, 573–580.

13. Toledano, M. B., Delaunay, A., Biteau, B., Spector, D., and Azevedo, D. (2003) Oxidative stress responses in yeast, in *Topics in Current Genetics. Yeast Stress Responses* (Hohmann, S., and Mager, P. W. H., eds.), Springer-Verlag, Berlin, Heidelberg, pp. 241–303.

14. Le Moan, N., Clement, G., Le Maout, S., Tacnet, F., and Toledano, M. B. (2006) The Saccharomyces cerevisiae proteome of oxidized protein thiols: contrasted functions for the thioredoxin and glutathione pathways. *J Biol Chem* **281**, 10420–10430.

15. Leichert, L. I. and Jakob, U. (2004) Protein thiol modifications visualized in vivo. *PLoS Biol* **2**, e333.

16. Baty, J. W., Hampton, M. B., and Winterbourn, C. C. (2005) Proteomic detection of hydrogen peroxide-sensitive thiol proteins in Jurkat cells. *Biochem J* **389**, 785–795.

17. Maeda, K., Finnie, C., and Svensson, B. (2004) Cy5 maleimide labelling for sensitive detection of free thiols in native protein extracts: identification of seed proteins targeted by barley thioredoxin h isoforms. *Biochem J* **378**, 497–507.

18. Jaffrey, S. R., and Snyder, S. H. (2001) The biotin switch method for the detection of S-nitrosylated proteins. *Sci. STKE*, **2001**, PL1.

19. Saurin, A. T., Neubert, H., Brennan, J. P., and Eaton, P. (2004) Widespread sulfenic acid formation in tissues in response to hydrogen peroxide. *Proc Natl Acad Sci USA* **101**, 17982–17987.

20. Poole, L. B., Zeng, B. B., Knaggs, S. A., Yakubu, M., and King, S. B. (2005) Synthesis of chemical probes to map sulfenic acid modifications on proteins. *Bioconjug Chem* **16**, 1624–1628.

21. Delaunay, A., Isnard, A. D., and Toledano, M. B. (2000) $H_2O_2$ sensing through oxidation of the Yap1 transcription factor. *EMBO J* **19**, 5157–5166.

22. Makmura, L., Hamann, M., Areopagita, A., Furuta, S., Munoz, A., and Momand, J. (2001) Development of a sensitive assay to detect reversibly oxidized protein cysteine sulfhydryl groups. *Antioxid Redox Signal* **3**, 1105–1118.

23. Mallis, R. J., Buss, J. E., and Thomas, J. A. (2001) Oxidative modification of H-ras: S-thiolation and S-nitrosylation of reactive cysteines. *Biochem. J* **355**, 145–153.

24. Delaunay, A., Pflieger, D., Barrault, M. B., Vinh, J., and Toledano, M. B. (2002) A thiol peroxidase is an $H_2O_2$ receptor and redox-transducer in gene activation. *Cell* **111**, 471–481.

25. Brennan, J. P., Wait, R., Begum, S., Bell, J. R., Dunn, M. J., and Eaton, P. (2004) Detection and mapping of widespread intermolecular protein disulfide formation during cardiac oxidative stress using proteomics with diagonal electrophoresis. *J Biol Chem* **279**, 41352–41360.

26. Cumming, R. C., Andon, N. L., Haynes, P. A., Park, M., Fischer, W. H., and Schubert, D. (2004) Protein disulfide bond formation in the cytoplasm during oxidative stress. *J Biol Chem* **279**, 21749–21758.

27. Chan, H. L., Gharbi, S., Gaffney, P. R., Cramer, R., Waterfield, M. D., and Timms, J. F. (2005) Proteomic analysis of redox- and ErbB2-dependent changes in mammary luminal epithelial cells using cysteine- and lysine-labelling two-dimensional difference gel electrophoresis. *Proteomics*, **5**, 2908–2926.

28. Marouga, R., David, S., and Hawkins, E. (2005) The development of the DIGE system: 2D fluorescence difference gel analysis technology. *Anal Bioanal Chem* **382**, 669–678.

# Chapter 14

# Analysis of Redox Relationships in the Plant Cell Cycle: Determinations of Ascorbate, Glutathione and Poly (ADPribose)Polymerase (PARP) in Plant Cell Cultures

Christine H. Foyer, Till K. Pellny, Vittoria Locato, and Laura De Gara

## Abstract

Reactive oxygen species (ROS) and low molecular weight antioxidants, such as glutathione and ascorbate, are powerful signaling molecules that participate in the control of plant growth and development, and modulate progression through the mitotic cell cycle. Enhanced reactive oxygen species accumulation or low levels of ascorbate or glutathione cause the cell cycle to arrest and halt progression especially through the G1 checkpoint. Plant cell suspension cultures have proved to be particularly useful tools for the study of cell cycle regulation. Here we provide effective and accurate methods for the measurement of changes in the cellular ascorbate and glutathione pools and the activities of related enzymes such poly (ADP-ribose) polymerase during mitosis and cell expansion, particularly in cell suspension cultures. These methods can be used in studies seeking to improve current understanding of the roles of redox controls on cell division and cell expansion.

**Keywords:** Antioxidants, plant cell suspension cultures, cell cycle, cellular redox state, oxidative stress, mitochondria, cell death, poly (ADPribose) polymerase.

## 1. Introduction

Cell cultures from different plant species such as tobacco (*Nicotiana tabacum*), tomato (*Lycopersicon esculentum*), carrot (*Daucus carota*), soybean (*Glycine max L.*), and *Arabidopsis thaliana* have been widely used to study cellular metabolism and gene expression. Such systems provide a homogenous cell source in which the uptake and processing of substrates, inhibitors, or effectors can be evaluated. However, it must be recognised that

John T. Hancock (ed.), *Methods in Molecular Biology, Redox-Mediated Signal Transduction, vol. 476*
© 2008 Humana Press, a part of Springer Science+Business Media, Totowa, NJ
DOI: 10.1007/978-1-59745-129-1_14

plant cell cultures represent a unique system, where cell-to-cell communication and signaling and other processes may differ from these occuring in whole tissues and organs. The suspension media contain all the required nutrients and elements that allow optimal growth and division while maintaining the cells in an undifferentiated state. It is important that the culture consists largely of single cells that are equally distributed throughout the liquid media and, hence, the suspension has to shaken during growth to disperse cell aggregates and prevent the formation clumped cells. The cells will then continuously grow until either one of the factors becomes limiting, causing cell growth to slow or division stops due to endogenous genetic controls that arrest the cell cycle.

The mitotic cell cycle can be readily synchonized in cell suspension cultures by the addition of aphidicolin (1). Hence, cell suspension cultures, particularly the tobacco BY-2 (Bright-Yellow 2 *N. tabacum* cv) (2) and the *Arabidopsis* cell suspension cultures (3) have been widely adopted for cell cycle studies (4, 5). The addition of the pro-oxidant, menadione to synchronized tobacco BY-2 cells causes specific arrests of the G1/S and the G2/M transitions in the cell cycle and, hence, slows DNA replication and mitosis (6). The oxidant-induced arrests of the cycle at the G1 and G2 phase checkpoints are accompanied by activation of defense genes (6). However, not all phases of the cell cycle are equally sensitive to oxidative arrest. For example, the G1 phase appears to be a more oxidation-sensitive checkpoint than the S-phase (6).

A well-established heterotrophic *Arabidopsis* cell suspension culture (3) has been the subject of several investigations into the responses of plant cells to oxidative stress. For example, this cell culture line has been used to analyze the affects of cellular oxidation on primary and secondary metabolism (7, 8) and on gene expression (9). The application of $H_2O_2$ to cultured cells (10) provokes a transcriptional response that mimics both biotic and abiotic stress responses, including expression of genes involved the hypersensitive response, programmed cell death and in ethylene and jasmonic acid signaling (11).

Reduction-oxidation (redox) reactions are central to cellular energy metabolism (12, 13). They are important regulators of plant growth, development, and defense (12, 13). Reactive oxygen species (ROS) such as $H_2O_2$ are formed during the reduction of molecular oxygen or water oxidation during metabolism (14). ROS are produced by the electron transport systems of chloroplasts and mitochondria and also by a variety of enzymes and redox reactions in almost every compartment of the plant cell. The accumulation of ROS in any tissue or cellular compartment is tightly controlled by the endogenous antioxidant systems, in which ascorbate and glutathione (GSH) are central components.

Cell suspension cultures have not only been used to study the effects of ROS and pro-oxidants that enhance ROS production but they have also proved to be useful in the study of the synthesis and roles of GSH and ascorbate. For example, cell suspension cultures can be easily supplied with a range of potential biosynthetic precursors, analogues, and inhibitors. This approach was used to investigate the pathway of ascorbate synthesis in *Arabidopsis* cultured cells *(15)*. The data obtained from the studies on cultured cells not only provided evidence in support of the L-galactose pathway of ascorbate biosynthetic described by Wheeler et al. *(16)*, but also showed that the *Arabidopsis* cells were also able to use two other routes of ascorbate production *(15)*. Methyl jasmonate treatment of *Arabidopsis* and tobacco BY-2 suspension cells increased the transcription of two late methyl jasmonate-responsive genes encoding enzymes of ascorbate synthesis *(17)*. The observed enhanced abundance of transcripts encoding GDP-mannose 3″, 5″-epimerase and a putative L-gulono-1, 4-lactone dehydrogenase/oxidase accompanied an increased rate of de novo ascorbate synthesis of the suspension cells *(17)*. Similarly, the apoplastic pathway of ascorbate degradation was recently described using cultured *Rosa* sp. ('Paul's Scarlet' rose) cell-suspension cultures *(18)*. This study identified several novel intermediates including 4-O-oxalyl-L-threonate in the degradation of ascorbate to oxalate and l-threonate *(18)*.

*Arabidopsis* cell cultures have also used to investigate the regulation of reduced GSH synthesis and provided evidence for the stimulatory role of cellular oxidation in increasing the abundance of this antioxidant *(3)*. Studies on cell suspension cultures have provided evidence that depletion of cellular antioxidants GSH and ascorbate cause inhibition of mitosis at G1/S transition. Buthionine-(S,R)-sulfoximine (BSO) is a nontoxic and highly specific inhibitor of γ-glutamyl cysteine synthetase, the first enzyme of the glutathione biosynthesis pathway *(19, 3)* that is often used to deplete cellular GSH pools *in planta*. When BSO was added to tobacco BY-2 suspension cultures, the cells arrested at the G1/S transition of the cell cycle *(20)*. The GSH-dependent developmental pathway was thus shown to be essential for initiation and maintenance of cell division during postembryonic root development *(20)*.

Studies on cell suspension cultures have demonstrate that oxidation of ascorbate (to dehydroascorbate) inhibits the progression through the cell cycle *(21–24)*. Moreover, the observation that depletion of GSH together with addition of dehydroascorbate caused additive affects on cell division in synchronized tobacco BY-2 suspension cultures led to the concept that these redox metabolites act in independent pathways to regulate the cell cycle *(25, 26)*. Other evidence in support of this view comes from the *Arabidopsis thaliana rml1* mutant, which is deficient in glutathione and cannot maintain a root meristem *(27)*. The addition of gamma-glutamyl cysteine or GSH, but not ascorbate or other reducing agents, restores root growth in the *rml1* mutant *(27)*. Hence, GSH has specific affects on cell division in the root meristem *(20)*. The addition of GSH to the BY-2 suspension cultures caused a slight inhibition rather than a stimulation of the cell cycle suggesting that the redox-mediated regulated progression through the cell cycle by ROS, ascorbate, and GSH is complex with each metabolite differentially affecting specific processes *(26)*.

Studies on the regulation of the cell cycle in suspension cultures have been pivotal in advancing current concepts of the respective roles for ROS, ascorbate, and glutathione in plant growth and development. ROS production, for example, is important in the orchestration of root growth and architecture *(28)* and the presence of high levels of dehydroascorbate has been linked to the formation of the population of slowly dividing cells in the root meristem called the quiescent center (QC), which is controlled and maintained by auxin *(29)* and modulation of mitochondrial function *(30)*. The relative amounts of reduced ascorbate and GSH pools are depleted in the QC such that the QC cells are maintained in a much more oxidized state than those of surrounding cells *(29)*. When the root cap is removed and the auxin signal is decreased the ascorbate and glutathione pools in the QC become more reduced and the cells start to divide *(29)*. This appears to be important factor in controlling the oxidized state of the QC cells in the root tip, the formation and maintenance of the QC being an inevitable developmental consequence of the oxidized microenvironment and concomitant alterations in mitochondrial activity and function *(29, 30)*. The synthesis of ascorbate and its presence in the reduced form in large quantities is important in embryo development and seedling growth *(31–33)*. Furthermore, apoplastic reactions involving ROS, ascorbate and monodehydroascorbate are important in the control of extension growth through regulation of cell wall cleavage, peroxidases ,and other enzymes involved in cell wall synthesis *(34–37)*.

Large changes in ascorbate contents are observed during the different growth stages of cells in culture suggesting that extensive turnover of the ascorbate pool occurs from the onset of the logarithmic growth phase to the pint where cell growth ceases *(24)*. In a recent study of the changes in the metabolite profile of *Arabidopsis* cells, Baxter et al. *(8)* reported an immediate (within the first hour) marked depletion of the total ascorbate pool, which they suggested may indicate the presence of a degree of oxidative stress even in the control cells, related to cell handling. This is rather an unusual observation because other studies on tobacco BY2, Arabidopsis, and sunflower cultures have observed an increase rather than a decrease in cellular ascorbate early in the growth phase *(24)* (Caretto S., Paradiso A., and De Gara L. unpublished results). Treatment of the *Arabidopsis* cells with the pro-oxidant, menadione resulted in a prolonged decrease in the cellular ascorbate content consistent with an increase in cellular redox state. Moreover, the accumulation of threonate in these conditions could result from ascorbate breakdown *(18)*. Menadione treatment also caused inhibition of the tricarboxylic acid cycle and of amino acid metabolism in the *Arabidopsis* cell suspension cultures, with diversion of carbon into the oxidative pentose phosphate pathway, a similar situation to that observed in the cells of the QC, which are maintained in an oxidised state *(30)*.

The aforementioned studies indicate the complexities of the redox signaling network that is involved in the orchestration of plant growth and development, as each ROS form (superoxide, $H_2O_2$, singlet oxygen) and each low molecular weight antioxidant (ascorbate, glutathione, tocopherol) can have discrete effects on plant cell division and growth, and can regulate discrete processes and sets of genes. At present, this signaling network remains poorly resolved and characterized, but recent information suggests that singlet oxygen not only modulates distinct sets of genes from those regulated by superoxide and $H_2O_2$ but these different ROS forms probably act antagonistically in the orchestration of genetically programmed cell suicide events *(38)*. Specific oxidation of target proteins and other signal molecules may be an intrinsic mechanism underpinning the perception and responses to environmental and developmental triggers. The recognition of damaged DNA by the enzyme poly(ADPribose) polymerase (PARP) has been linked to cell signalling events, initiating structural changes and isoform replacement and to protection against a range of abiotic stresses *(39)*.

Although cellular oxidation is still considered by many to be a largely negative event, the probability of lasting damage arising from enhanced ROS production/accumulation is largely one of degree or threshold. BY-2 cell suspensions show a dose-dependent sensitivity to oxidative stress. Low concentrations caused transient inhibition of the cell cycle and enhanced defense, whereas high concentrations cause cell death *(6)*. The oxidative burst triggered by pathogens or elicitors such as harpin has been studied extensively in cell cultures *(9, 40–43)*. Decreases in transcripts encoding defence genes associated with antioxidant metabolism and ROS scavenging are observed early in the hypersensitive cell death response, suggesting that redox signaling pathways are important in the systemic acquired resistance and programmed cell death (PCD) responses to pathogens *(44, 45)*. Moreover, a strong correlation has been found between ascorbate depletion, decreased ascorbate peroxidase activities and PCD responses in tobacco BY cultured cells *(46, 47)*. Similarly, the heat stress-induced PCD response in tobacco BY-2 cell cultures can be inhibited by ascorbate *(48)*. Barley aleurone and wheat endosperm cells, which undergo PCD at the beginning of germination and during kernel development respectively, remain alive longer when they are submitted to treatments that increase the cellular antioxidant capacity *(49)* (Paradiso A. and De Gara L., unpublished).

Studies on the roles of redox regulation on the mitotic cell cycle and other key processes in plant cell cultures requires not only accurate and effective methods for ROS detection but also similar methods for the assay of antioxidant pools and related enzymes. In such studies samples usually consisting of mass of cells are harvested throughout the growth cycle for analysis. Cellular antioxidants are very susceptible to oxidation and degradation during extraction and hence metabolism has to be rapidly arrested in order to preserve the *in planta* redox state of metabolites and proteins. The series of procedures and methods that are described below are central to the analysis of the cellular redox state and the impact of changes in redox metabolism on metabolic integration and oxidative signalling in plant cell cultures. They are also important in any consideration of how ascorbate or glutathione influences cell division and controls the plant cell cycle in cells in suspension culture.

## 2. Materials

1. 1X MS media with vitamins containing 3% sucrose, 0.5 mg/L NAA and 0.05 mg/L kinetin, pH 5.8.
2. Trypan Blue Stain 0.4% (Gibco, cat. no. 15250-061).
3. 14-mL Falcon tubes.
4. 1X phosphate-buffered saline (PBS).
5. Nylon membrane filters (0.45-μm pore size; Whatman, cat. no. 7404004).
6. Perchloric acid ($HClO_4$).
7. 120 m$M$ sodium phosphate buffer (pH 5.6).
8. 2.5 $M$ $K_2CO_3$.
9. pH paper.
10. 120 m$M$ sodium phosphate buffer (pH 7.6).
11. Ascorbate oxidase.
12. 20 m$M$ dithiothreitol (DTT).
13. 2x reaction buffer (240 m$M$ sodium phosphate buffer with 12 m$M$ ethylene diamine tetraacetic acid [EDTA] at pH 7.5).
14. 12 m$M$ 5, 5′-dithiobis-(2-nitrobenzoic acid), (DTNB).
15. Glutathione reductase.
16. 10 m$M$ NADPH.

17. 2-vinyl pyridine.

18. 1 m$M$ EDTA (pH 7.6).

19. 0.15% (v/v) Triton X-100.

20. Extraction buffer (30 m$M$ 3-($N$-Morpholino)-propanesulphonic acid [MOPS], 1 mM phenylmethanesulphonylfluoride [PMSF], 2 m$M$ EDTA, pH 7.5).

21. 50 m$M$ TRIS-HCl (pH 8.0), 60 µ$M$ cytochrome $c$, 3 m$M$ galactono lactone, 1 m$M$ KCN.

22. 35 m$M$ ascorbate.

23. Protoplast extraction buffer (0.4 $M$ mannitol, 20 m$M$ 2-Morpholinoethanesulfonic acid, monohydrate (MES), pH 5.5 with HCl, with 0.25% (w/v) cellulase, 0.05% (w/v) pectolyase, 0.1% (v/v) pectinase; added on the day).

24. CelLytic P Kit (Sigma C2360) containing 4X NIB (Nuclear Isolation Buffer), 2.3 $M$ sucrose stock, and Nuclei Extraction Buffer.

25. Protease inhibitor cocktail (Sigma P9599).

26. Dounce homogenizer.

27. Vortex with tube attachment.

28. 1X TBS buffer (Tris-buffered saline: Tris-HCl 20 m$M$, pH 7.5, containing 150 m$M$ NaCl).

29. Dried skimmed milk (*see* **Note 6**).

30. 0.1% (v/v) Tween 20.

31. Anti-rabbit IgG alkaline phosphatase conjugate.

32. Desktop scanner and BioRad QuantityOne program.

## 3. Methods

### 3.1. Cell Cultures

#### 3.1.1. Growth of Suspension Cultures

*Arabidopsis thaliana* Landsberg *erecta* cultures MM1 and MM2d can be maintained as described previously *(3)*.

1. Weekly dilute 3.5 mL of saturated culture in 100 mL of fresh 1X MS media with vitamins containing 3% sucrose, 0.5 mg/L NAA and 0.05 mg/L kinetin, pH 5.8 (*see* **Note 1**).

2. Shake cultures are in 250-mL Erlenmeyer flasks at a rotation of 120 rpm. MM1 is cultured for 16-h light (~10 µmol/m² PAR)/8 h dark at 22°C. MM2d is cultured constantly in the dark at 25°C.

3. Follow the growth of the culture by measuring the optical density at 595 nm for at least six replicate aliquots of 250 µL using a plate reader spectrophotometer. For high concentrations,

the culture should be diluted one in four in 1x PBS buffer prior to measurement.

*3.1.2. Cell Viability*

The viability of the cells is tested using Trypan Blue Staining.

1. Gently mix aliquots of the cultures with equal volume of Trypan Blue Stain 0.4%.

2. Observe under a light microscope. Dead cells show up blue due to the penetration of the dye through disrupted membranes.

*3.1.3. Harvest*

1. Using a pipet place cells into 2-mL microfuge tubes with an insertion in the bottom (*see* **Note 2**).

2. Centrifugate in 14-mL Falcon tubes at $250g$ for 2 min with no brake applied.

3. Then snap freeze the pellet of packed cells in the microfuge tube in liquid nitrogen and stored at $-80°C$.

4. Alternatively, cells can be immediately re-suspended in the appropriate buffer for analysis.

This method can be also used to determine the cell fresh and dry weights of known culture volumes. The medium that is allowed to escape through the microfuge insertion tube (*see* **Note 2**) can be used to analyze the media as well as to estimate the packed cell volume by subtraction of the recovered media volume to the cell suspension volume.

5. Otherwise, larger volumes cells are harvested by vacuum aspiration on nylon membrane filters (0.45-µm pore size).

***3.2. Metabolite Analysis***

*3.2.1. Extraction of Metabolites*

1. Resuspend the harvested samples (from Sect. 3.1.3.) in a minimum of 550 µL of 1 *M* perchloric acid ($HClO_4$).

2. Freeze in liquid nitrogen.

3. Thaw and vortex the frozen samples.

4. Clarify the homogenate in a cooled microfuge (at $4°C$) for 10 min at maximum speed.

*3.2.2. Adjustment of pH Before Assay*

1. Transfer the supernatant (500 µL) to fresh microtubes containing 120 m*M* sodium phosphate buffer (100 µL; pH 7.6).

2. Adjust the pH of each sample carefully to pH values between pH 5.0 and 6.0 for ascorbate assays by the addition of 2.5 *M* $K_2CO_3$. Attention: do not overshoot the pH as ascorbate is rapidly oxidized at alkaline pH values! For glutathione determinations alone, the pH can be adjusted to pH 7.0 and remain relatively stable. Make a note of the added volume and check the pH with pH paper. Care! The reaction is volatile!

3. Remove the insoluble $KClO_4$ produced during adjustment of the pH is by centrifugation at $4°C$ and maximum speed

for 10 min. The supernatant is used for the assay of metabolites such as ascorbate and glutathione.

Reduced ascorbate is measured by the change in absorption at 265 nm ($OD_{265}$) as ascorbate is depleted from the reaction mixtures by oxidation to dehydroascorbate.

1. Established a stable background reading for the extract (200 µL at pH 5.6) in 120 m$M$ sodium phosphate buffer (800 µL at pH 5.6).

2. Add ascorbate oxidase (0.25 U in 5 µL of reaction buffer [*see* **Note 3**]).

3. Follow the reaction is until a minimum is reached (usually less than 1 min).

To measure total ascorbate, dehydroascorbate is first reduced to ascorbate in the extracts.

4. Incubate each sample of extract (100 µL at pH 5.6) with Dithiothreitol (DTT; 10 µL 20 m$M$) and 120 m$M$ sodium phosphate buffer (pH 7.6; 140 µL 120 m$M$) for up to 30 min.

5. Stop this reaction by adding 120 m$M$ sodium phosphate buffer (750 µL; pH 5.6).

6. Measure the total ascorbate by the addition of ascorbate oxidase as previously described (*see* **step 2**).

Ascorbate concentrations are calculated from the $OD_{265}$ change and the extinction coefficient for ascorbate at this wavelength (12.8/mM/cm). The amount of oxidized ascorbate is equivalent to the difference between the total and reduced ascorbate. Ascorbate concentrations can be confirmed by production of standard curves with known amounts of pure ascorbate. Recovery experiments involving mixtures of extract and known amounts of ascorbate should be performed at the outset for each new cell type to be studied.

Total glutathione (GSH plus GSSG) can be analyzed with a microplate-based cycling assay procedure. In this method, reduced GSH is measured by the interaction with dithio-bis-2-nitrobenzoic acid (DTNB) forming 2-nitro-5-thiobenzoic acid with an absorbance peak at 412 nm. Glutathione reductase (GR) is used to reduce oxidized glutathione (GSSG) to GSH in the presence of NADPH.

1. Per reaction, mix 100 µL of the 2x reaction buffer (240 m$M$ sodium phosphate buffer with 12 m$M$ EDTA at pH 7.5) with 10 µL of 12 m$M$ DTNB and 0.25 U GR (*see* **Note 3**).

2. Make up to a volume of 180 µL with ddH$_2$O.

3. Add 10 µL of 10 m$M$ NADPH and 10 µL of the extract (pH 6.0–7.0) to the flat-based microplates.

4. With a multichannel pipet, add 180 µL of the reaction mixture.

5. Monitored the $OD_{412}$ change over 10 min.

The initial rate of this reaction correlates to the total glutathione content and is compared to standard curves for up to 0.3 nmol GSH (*see* **ref. 51**). GSSG is quantified in a similar manner but only following prior removal of GSH from the extracts with 2-vinyl pyridine. GSH forms an insoluble complex with 2-vinyl pyridine and can therefore be removed easily from the extract.

1. To the neutralized metabolite extract (200 µL), add 5 µL of 2-vinyl pyridine.

2. Gently mixed the sample and incubate at room temperature for 30 min.

3. Then, transfer to ice for 30 min.

4. Subject the mixture to centrifugation for 10 min at 4°C with maximum speed.

5. Avoid removing any 2-vinyl pyridine with the sample for assay as far as possible, as this compound will inhibit the reaction. The supernatant is then used for GSSG determination in the cycling assay, as described above.

6. GSH and GSSG concentrations should be confirmed by production of standard curves with known amounts of pure GSH or GSSG. Please note that standard curves for GSSG should include the 2-vinyl pyridine extraction step. As with ascorbate, recovery experiments, as illustrated in **Fig.14.1**, involving mixtures of extract and known amounts of GSH and/or GSSG should be performed at the outset for each new cell type to be studied.

*3.2.5. Extraction and Assay of Total Protein in Acid Precipitates*

1. Solubilize proteins precipitated in the pellet after acid extraction using 500 µL of 120 m$M$ sodium phosphate buffer with 1 m$M$ EDTA (pH 7.6).

2. Extract the pellet on ice using a microfuge pestle.

3. Centrifuge the sample (10 min; 4°C; max. speed).

4. Determine the soluble proteins using a standard Bradford assay.

**3.3. Enzyme Activity Measurements**

*3.3.1. Extraction Procedures*

1. Homogenize samples to a fine powder in liquid nitrogen.

2. For the extraction of l-galactono-1,4-lactone dehydrogenase (GLDH), add 0.15% (v/v) Triton X-100 to the extraction buffer (30 m$M$ MOPS, 1 m$M$ PMSF, 2 m$M$ EDTA, pH 7.5).

3. Homogenized samples on ice until fully thawed.

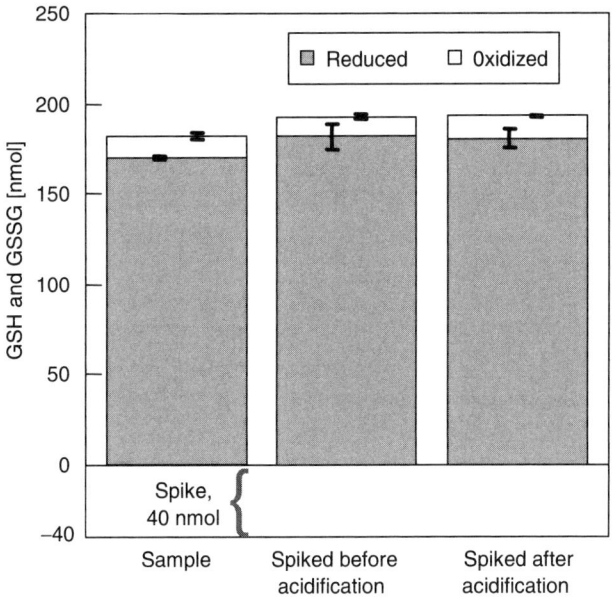

Fig. 14.1. Recovery of glutathione (GSH). Replicate samples from a stationary cell culture were harvested as described in the Methods. A known amount of pure reduced glutathione was added to subsamples (spike) either to the cell pellet before or after the addition of perchloric acid and all samples analysed in parallel. Amounts of standard pure GSH recovered ranged from 90% to 110%. Values less than 80% and greater than 130% are unacceptable. The amounts of oxidized GSH are very similar in all experiments showing that GSH is not oxidized during the extraction procedures. Oxidation of GSH during extraction must be minimal.

4. Subject the extract to centrifugation (15 min, 4°C, 18,000$g$), and the supernatant fractions are used for the following assay procedures.

5. Determine the protein content of each supernatant fraction using a standard Bradford assay.

GLDH is bound to the inner mitochondrial membrane and Triton X-100 can be added if required to liberate the enzyme into the soluble supernatant fraction. Its inclusion is not necessary when ascorbate oxidase is assayed, even though this enzyme is bound to the cell wall. Triton X-100 also interferes with the following assays and should be avoided. High salt can be used to liberate cell wall-bond ascorbate oxidase (37).

*3.3.2. Galactono-1, 4-Lactone Dehydrogenase Activity*

GLDH (EC 1.3.2.3) activity is assayed by the coupled reduction of cytochrome $c$, which leads to a specific change in OD$_{550}$ as described by Bartoli et al. (52).

1. Make the reaction buffer up fresh daily with 50 m$M$ TRIS-HCl (pH 8.0), 60 µ$M$ cytochrome $c$, 3 m$M$ galactono lactone (*see* **Note** 4), and 1 m$M$ KCN to inhibit the reoxidation of the reduced cytochrome $c$ (**Fig.14.2**).

2. Initiate the reaction by the addition of supernatant sample.

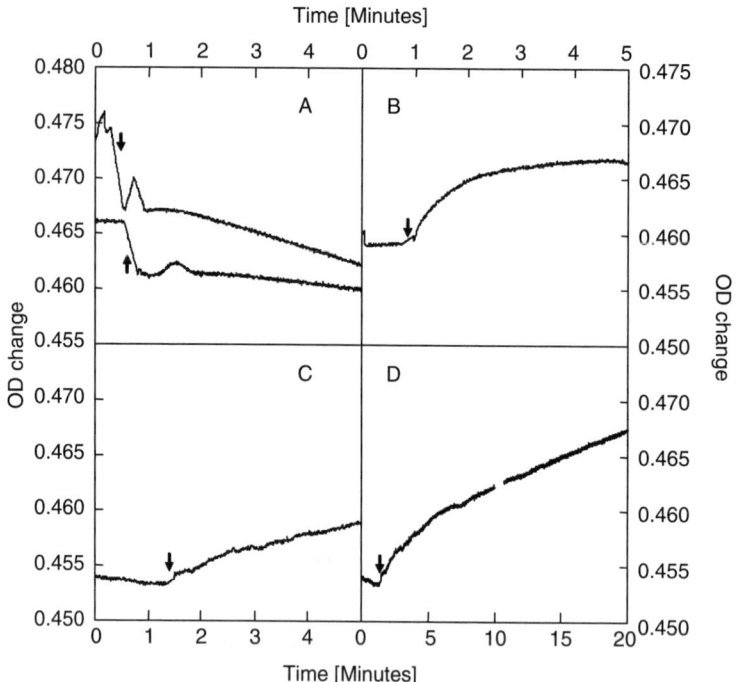

Fig. 14.2. Optimisation of GLDH assay. The presence of Triton X100 in the enzyme reaction mixtures causes artefacts. Although some literature methods use extraction procedures with Triton X100, we do not recommend its use in this assay. The presence of Triton X100 in the reaction mixtures leads to changes in optical density (**A**), which are not observed when Triton X100 is omitted from the reaction (**B, C**). The addition of KCN, which inhibits the reoxidation of the reduced cytochrome c, enables a steady-state reaction rate to be reached (**B**). In the presence of KCN the reaction is stable for a long period (**C, D**). The reaction is stable for a long period (**C**). We recommend that the reaction rate is measured over a 10-min period (**D**). The arrow indicates the addition of the extract, (equivalent to 50 mg protein).

3. Measured the enzyme activity by the change in absorption at 550 nm ($OD_{550}$).

4. The reaction rate should be followed for at least 10 min to ensure linearity.

5. GLDH activity is then calculated from that change in reduced cytochrome $c$ content with time. The extinction coefficient of reduced cytochrome $c$ is 21/mM/cm under these conditions.

*3.3.3. Ascorbate Oxidase Activity*

Ascorbate oxidase (EC 1.10.3.3) is determined using a modification of the procedures employed for measuring ascorbate.

1. Add an aliquot of supernatant to 120 m$M$ phosphate buffer (pH 5.6) containing 6 m$M$ EDTA in a final volume of 1 mL.

2. When a stable reading has been obtained, initiate the reaction by the addition of 8.5 μL of 35 m$M$ ascorbate.

3. Measured the enzyme activity by the change in absorption at 290 nm ($OD_{290}$). The decrease in absorbance at $OD_{290}$ is followed as ascorbate is oxidized.

4. Calculate the rate of ascorbate oxidation with time using the extinction coefficient of ascorbate (2.7 m$M$/cm at 290 nm).

**3.3.4. Poly(ADPribose) Polymerase**

To isolate nuclei, protoplasts are obtained from *Arabidopsis thaliana* cell culture.

**3.3.4.1. Preparation of Protoplasts**

1. Resuspend circa 500 mg (FW) *Arabidopsis* cells in 2 mL of protoplast extraction buffer (0.4 $M$ mannitol, 20 m$M$ MES, pH 5.5 with HCl with 0.25% (w/v) cellulase, 0.05% (w/v) pectolyase, 0.1% (v/v) pectinase; added on the day).
2. Incubated the cells in the dark with gentle agitation for about 1 h to obtain protoplasts.
3. Monitor progress by observing sub-samples of the cells under the light microscope.
4. Recover protoplasts by centrifugation at 1,000$g$ for 3 min at room temperature without the brake.
5. Wash the pellet twice with the protoplast extraction buffer without enzymes (3 mL) using the same centrifugation conditions.

**3.3.4.2. Isolation of Nuclei**

The isolation of nuclei is based on the CelLytic P Kit. Each operation should be performed on ice.

1. Resuspend protoplasts in 2 mL of NIBA (1XNuclear Isolation buffer with 1% (v/v) Protease Inhibitor Cocktail).
2. Disrupt using a Dounce homogenizer, with ten turns with the loose fitting pestle followed by ten turns with the tight fitting pestle.
3. Subsequently lyse the cells by incubation on ice after the addition of 0.3% (v/v) Triton X-100.
4. Monitor progress under the light microscope.

**3.3.4.3. Purification of Nuclei**

1. Carefully apply aliquots of the lysate (500 µL) to the top of 800 µL of a 1.5 $M$ sucrose solution (obtained by diluting a 2.3 $M$ sucrose stock with 1X NIB buffer).
2. Centrifuge the tubes for 10 min at 4°C, 12,000$g$ without the brake.
3. Wash the pellets twice in 1 mL of NIBA and then combined in 100 µL of Extraction Buffer for Nuclei containing 1% (v/v) Protease Inhibitor Cocktail and 5 m$M$ DTT.

**3.3.4.4. Extraction of Nuclear Proteins**

1. Mix the nuclear suspension at medium-high speed for 30 min at 4°C in a vortex with tube attachment.
2. Centrifuge for 10 min at 4°C; 12,000$g$.
3. Recover the supernatant.
4. Freeze in liquid nitrogen and store at –80°C.

**3.3.4.5. Immuno-Dot Blot Assay for Poly (ADPribose) Polymerase Activity**

The activity of PARP is measured by the immuno-detection of the Poly (ADP)ribose chain as described by De Block et al. *(39)*.

1. Determine the total concentration of the extracted nuclear proteins with a standard Bradford assay.

2. Normalize the samples to a concentration of 0.5 mg protein in a volume of 15 μL with 1X TBS buffer.

3. Spot the samples are on to a Hybond C membrane, which has been presoaked in TBS buffer (*see* **Note** 5), and air-dried.

4. Float the membrane is on TBS buffer until evenly wet followed by submergence.

5. Follow this by incubation in TBS with 5% (w/v) dried skimmed milk (*see* **Note** 6) and 0.1% (v/v) Tween 20 with gentle agitation for 1 h.

6. Dilute the anti-PAR primary antibody 1:2,500 in TBS with 1% (w/v) dried skimmed milk and 0.1% (v/v) Tween 20.

7. Incubate the membrane in this for 1 h at room temperature with gentle agitation.

8. Then wash the membrane five times for 5 min in TBS with 1% (w/v) dried skimmed milk and 0.1% (v/v) Tween 20.

9. Follow by incubation with the anti-rabbit IgG alkaline phosphatase conjugate (diluted 1:5,000 in 1% blocking solution) for 1 h at room temperature with gentle agitation.

10. Wash five times for 5 min in TBS with 0.1% (v/v) Tween 20, followed by two rinses with TBS.

11. Develop the membrane by incubation in BCIP-NBT for 10–30 min until a clear signal is obtained.

12. Stop the reaction is by washing the membrane repeatedly with ddH$_2$O.

13. Capture the image as a TIFF document, using a desktop scanner, and analyse spot intensities with the BioRad Quantity-One program.

## 4. Notes

1. Dissolve the phytohormones in 1 $N$ NaOH (approx 1 mL for 10 mg) and make up 1 mg/mL stock by slowly adding ddH$_2$O. Add phytohormones before autoclaving and adjust the pH to 5.6 with KOH.

2. Use a 2-mL microfuge tube slit with the very top of hypodermic needle. Do not pierce it fully as this will create a hole through which cells will escape.

3. Ascorbate oxidase from *Cucurbita sp.* (Sigma A0157) lyophilized powder. Make up a 50 U/mL stock solution in 120 m$M$ sodium phosphate buffer with 1 m$M$ EDTA (pH 5.6). Either store at 4°C and use within 1 wk of preparation, or store in aliquots at −20°C. The enzyme is inactivated by freeze-thawing so do not freeze and thaw repeatedly. Glutathione reductase from baker's yeast (*S. cerevisiae*) (Sigma; G3664) is diluted in reaction buffer to a concentration of 10 U/mL and stored in aliquots at −20°C.

4. It is difficult to obtain stocks of pure L(+) Galactono γ lactone commercially at present. Sigma no longer stocks this substrate. Hence, we currently rely on existing stocks of the pure metabolite.

5. Avoid using phosphate-buffered saline buffer in conjunction with alkaline phosphatase detection.

6. The anti-PAR antibody shows cross reactions with bovine serum albumin, do not use it as a blocking agent.

# References

1. Menges, M. and Murray, J. A. (2002) Synchronous *Arabidopsis* suspension cultures for analysis of cell-cycle gene activity. *Plant J* **2**, 203–212.

2. Nagata, T., Nemoto, Y., and Hasezawa, S. (1992). Tobacco BY-2 cell line as the 'HeLa' cell in the cell biology of higher plants. *Int Rev Cytol* **132**, 1–30.

3. May, M. J. and Leaver, C. J. (1993). Oxidative stimulation of glutathione synthesis in *Arabidopsis thaliana* suspension cultures. *Plant Physiol* **103**, 621–627.

4. Meijer, M. and Murray, J. A.H. (2001) Cell cycle controls and the development of plant form. *Curr Opin Plant Biol* **4**, 44–49.

5. Dewitte, W. and Murray, J. A.H. (2003) The plant cell cycle. *Ann Rev Plant Biol* **54**, 235–264.

6. Riechheld, J. P., Vernoux, T., Lardon, F., Vanmontagu, M., and Inze, D. (1999) Specific checkpoints regulate cell cycle progression in response to oxidative stress. *Plant J* **17**, 647–656.

7. Tiwari, B. S., Belenghi, B., and Levine, A. (2002) Oxidative stress increased respiration and generation of reactive oxygen species, resulting in ATP depletion, opening of mitochondrial permeability transition, and programmed cell death. *Plant Physiol* **128**, 1271–1281.

8. Baxter, C. J., Redestig, H., Schauer, N., Repsilber, D., Patil, K. R., Nielsen, J., Selbig, J., Liu, J., Fernie, A. R., and Sweetlove, L. J. (2007) The metabolic response of heterotrophic *Arabidopsis* cells to oxidative stress. *Plant Physiol* **143**, 312–325.

9. Desikan, R., Reynolds, A., Hancock, J. T., and Neill, S. J. (1998) Harpin and hydrogen peroxide both initiate programmed cell death but have differential effects on defence gene expression in *Arabidopsis* suspension cultures. *Biochem J* **330**, 115–120.

10. Desikan, R., A.-H.-Mackerness, S., Hancock, J. T., and Neill, S. J. (2001) Regulation of the *Arabidopsis* transcriptome by oxidative stress. *Plant Physiol* **127**, 159–172.

11. Vandenabeele, S., Van Der Kelen, K., Dat, J., Gadjev, I., Boonefaes, T., Morsa, S., Rottiers, P., Slooten, L., Van Montagu, M., Zabeau, M., Inze, D., and Van Breusegem, F. (2002) A comprehensive analysis of hydrogen peroxide-induced gene expression in tobacco. *Proc Natl Acad Sci USA* **100**, 16113–16118.

12. Foyer, C. H. and Noctor, G. (2005) Oxidant and antioxidant signalling in plants:

a re-evaluation of the concept of oxidative stress in a physiological context. *Plant Cell Environ* **28**, 1056–1071.

13. Foyer, C. H. and Noctor, G. (2005) Redox homeostasis and antioxidant signalling: a metabolic interface between stress perception and physiological responses. *Plant Cell* **17**, 1866–1875.

14. Foyer, C. H. and Noctor, G. (2000). Oxygen processing in photosynthesis: regulation and signalling. *New Phytol* **146**, 359–388.

15. Davey, M. W., Gilot, C., Persiau, G., Østergaard, J., Han, Y., Bauw, G. C., and Van Montagu, M. C. (1999) Ascorbate biosynthesis in *Arabidopsis* cell suspension culture. *Plant Physiol* **121**, 535–544.

16. Wheeler, G. L., Jones, M. A., and Smirnoff, N. (1998). The biosynthetic pathway of vitamin C in higher plants. *Nature* **393**, 365–369.

17. Wolucka, B. A., Goossens, A., and Inzé, D. (2005) Methyl jasmonate stimulates the *de novo* biosynthesis of vitamin C in plant cell suspensions. *J Exp Bot* **56**, 2527–2538.

18. Green, M. A. and Fry, S. C. (2004) Degradation of vitamin C in plant cells via enzymic hydrolysis of 4-*O*-oxalyl-l-threonate. *Nature* **433**, 83–87.

19. Griffith, O. W. and Meister, A. (1979) Potent and specific inhibition of glutathione synthesis by buthionine sulfoximine (S-n butyl homocysteine sulfoximine). *J Biol Chem* **254**, 7558–7560.

20. Vernoux, T., Wilson, R. C., Seeley, K. A., Reichheld, J. P., Muroy, S., Brown, S., Maughan, S. C., Cobbett, C. S., Van Montagu, M., Inzé, D., May, M. J., and Sung Z. R. (2000) The ROOT MERISTEMLESS1/CADMIUM SENSITIVE2 gene defines a glutathione-dependent pathway involved in initiation and maintenance of cell division during postembryonic root development. *Plant Cell* **12**, 97–110.

21. Liso, R., Innocenti, A. M., Bitonti, M. B., and Arrigoni, O. (1988) Ascorbic acid – induced progression of quiescent centre cells from $G_1$ to S phase. *New Phytol* **110**, 469–471.

22. De Gara, L. and Tommasi, F. (1999) Ascorbate redox enzymes: a network of reactions involved in plant development. *Recent Res Dev Phytochem* **3**, 1–15.

23. de Pinto M. C., Francis D., and De Gara L. (1999) The redox state of the ascorbate-dehydroascorbate pair as a specific sensor of cell division in tobacco BY-2 cells. *Protoplasma* **209**, 90–97.

24. de Pinto, M. C., Tommasi, F., and De Gara, L., (2000) Enzymes of the ascorbate biosynthesis and ascorbate-glutathione cycle in cultured cells of tobacco Bright Yellow 2. *Plant Physiol Biochem* **38**, 541–550.

25. Potters, G., De Gara, L., Asard, H., and Horemans, N. (2002) Ascorbate and glutathione: guardians of the cell cycle, partners in crime? *Plant Physiol Biochem* **40**, 537–548.

26. Potters, G., Horemans, N., Bellone, S., Caubergs, R. J., Trost, P., Guisez, Y., and Asard, H. (2004) Dehydroascorbate influences the plant cell cycle through a glutathione-independent reduction mechanism. *Plant Physiol* **134**, 1479–1487.

27. Cheng, J.-C., Seeley, K. A., and Sung, Z. R. (1995) RML7 and RML2, *Arabidopsis* genes required for cell proliferation at the root tip. *Plant Physiol* **107**, 365–376.

28. Carol, R. J., Takeda, S., Linstead, P., Durrant, M. C., Kakesova, H., Derbyshire, P., Drea, S., Zarsky, V., and Dolan, L. (2005) A RhoGDP dissociation inhibitor spatially regulates growth in root hair cells. *Nature* **438**, 1013–1016.

29. Jiang, K., Meng, Y. L., and Feldman, L. J. (2003) Quiescent center formation in maize roots is associated withan auxin-regulated oxidising environment. *Development* **130**, 1429–1438.

30. Jiang, K., Ballinger, T., Li, D., Zhang, S., and Feldman, L. A. (2006) Role for mitochondria in the establishment and maintenance of the maize root quiescent center. *Plant Physiol* **140**, 1118–1125.

31. Bailly, C., Audigier, C., Ladonne, F., Wagner, M. H., Coste, F., Corbineau, F., and Come, D. (2001) Changes in oligosaccharide content and antioxidant enzyme activities in developing bean seeds as related to acquisition of drying tolerance and seed quality. *J Exp Bot* **52**, 701–708.

32. De Gara, L., de Pinto, M. C., Moliterni, V. M., and D'Egidio, M. G. (2003). Redox regulation and storage processes during maturation in kernels of *Triticum durum*. *J Exp Bot* **54**, 249–258.

33. Tommasi, F., Paciolla, C., de Pinto, M. C., and De Gara, L. (2001) A comparative study of glutathione and ascorbate metabolism during germination of *Pinus pinea* L. seeds. *J Exp Bot* **52**, 1647–1654.

34. Córdoba-Pedregosa, M. C., Cordoba, F. Villalba, J. M., and Gonzáles-Reyes J. A. (2003) Zonal changes in ascorbate and hydrogen peroxide contents, peroxidase, and ascorbate-related enzymes activities in onion roots. *Plant Physiol* **131**, 1–10.

35. de Pinto, M. C. and De Gara, L. (2004) Changes in the ascorbate metabolism of both apoplastic and symplastic spaces are involved in cell differentiation. *J Exp Bot* **55**, 2559–2569.

36. Dumville, J. C. and Fry, S. C. (2003) Solubilisation of tomato fruit pectins by ascorbate: a possible non-enzymic mechanism of fruit softening. *Planta* **217**, 951–961.

37. Pignocchi, C., Fletcher, J. E., Barnes, J., and Foyer, C. H. (2003) The function of ascorbate oxidase (AO) in tobacco ( *Nicotiana tabacum* L.). *Plant Physiol* **132**, 1631–1641.

38. Apel, K., and Hirt, H. (2004) Reactive oxygen species: metabolism, oxidative stress, and signal transduction. *Ann. Rev Plant Biol* **55**, 373–399.

39. De Block, M., Verduyn, C., De Brouwer, D., and Cornelissen, M. (2004) Generating stress tolerant crops by economizing energy consumption. *Pflanzenschutz-Nachrichten Bayer* **57**, 105–110.

40. Baker, C. J., Orlandi, E. W., and Mock, N. M. (1993) Harpin, an elicitor of hypersensitive response in tobacco caused by *Erwinia amylovora*, elicits active oxygen production in suspension cells. *Plant Physiol* **102**, 1341–1344

41. Andi, S., Taguchi, F., Toyoda, K., Shiraishi, T., and Ichinose, Y. (2001) Effect of methyl jasmonate on harpin-induced hypersensitive cell death, generation of hydrogen peroxide and expression of *PAL* mRNA in tobacco suspension cultured BY-2 cells. *Plant Cell Physiol* **42**, 446–449.

42. Popham, P., Pike, S., and Novacky, A. (1995) The effects of harpin from *Erwinia amylovora* on the plasmalemma of suspension cultured tobacco cells. *Physiol Mol Plant Pathol* **47**, 39–50.

43. Xie, Z. and Chen, Z. (2000) Harpin-induced hypersensitive cell death is associated with altered mitochondrial functions in tobacco cells. *Mol Plant-Microbe Interact* **13**, 183–190.

44. De Gara, L., de Pinto, M. C., and Tommasi, F. (2003) The antioxidant system vis à vis reactive oxygen species during plant pathogen interaction. *Plant Physiol Biochem* **41**, 863–870.

45. Mittler, R., Vanderauwera, S., Gollery, M., and Van Breusegem, F. (2004) Reactive oxygen gene network of plants. *Trends Plant Sci* **9**, 490–498.

46. de Pinto, M. C., Tommasi, F., and De Gara, L. (2002) Changes in the antioxidant systems as part of the signalling pathway responsible for the programmed cell death activated by nitric oxide and reactive oxygen species in tobacco BY-2 cells. *Plant Physiol* **130**, 698–708.

47. de Pinto, M. C., Paradiso, A., Leonetti P., and De Gara L., (2006) Hydrogen peroxide, nitric oxide and cytosolic ascorbate peroxidase at the crossroad between defence and cell death. *Plant J* **48**, 784–795.

48. Vacca, R. A., de Pinto, M. C., Valenti, D., Passerella, S., Marra, E., and De Gara, L. (2004) Reactive oxygen species production, impairment of glucose oxidation and cytosolic ascorbate peroxidase are early events in heat-shock induced programmed cell death in tobacco BY-2 cells. *Plant Physiol* **134**, 1100–1112.

49. Beligni, M. V., Fath, A., Bethke, P. C., Lamattina, L., and Jones, R. L. (2002) Nitric oxide acts as an antioxidant and delays programmed cell death in barley aleurone layers. *Plant Physiol* **129**, 1642–1650.

50. Foyer, C. H., Rowell, J., and Walker, D. A. (1983). Measurement of the ascorbate content of spinach leaf protoplasts and chloroplasts during illumination. *Planta* **157**, 239–244.

51. Noctor, G. and Foyer, C. H. (1998). Simultaneous measurement of foliar glutathione, γ-glutamyl cysteine and amino acids by high-performance liquid chromatography: comparison with two other assay methods for glutathione. *Anal Biochem* **264**, 98–110.

52. Bartoli C. G., Pastori, G. M., and Foyer, C. H. (2000). Ascorbate biosynthesis in mitochondria is linked to the electron transport chain between complex III and IV. *Plant Physiol* **123**: 335–343.

# Chapter 15

# Generation and Detection of *S*-Nitrosothiols

## Christian Lindermayr, Simone Sell, and Jörg Durner

## Abstract

Nitric oxide (NO) plays a pivotal role in cellular signaling in many different organisms as the result of the modification of protein activities/functions by protein S-nitrosylation. This NO-dependent post-translational modification is based on the attachment of NO to the sulfur moiety of cysteine residues. However, the instability of *S*-nitrosothiols makes it difficult to analyze this type of protein modification *in vitro* as well as *in vivo*. Jeffrey and colleagues developed a method—named the biotin switch method—that allows the detection and purification of *S*-nitrosylated proteins. The principle behind this technology is the substitution of the NO group by a biotin linker in a three-step procedure. First, the all free thiol groups are blocked with a thiol-reactive agent, followed by selective reduction of the *S*-nitrosylated cysteine residues using ascorbate. In the final step, the reduced thiol groups are labeled with a biotin linker, so that the previously *S*-nitrosylated cysteine residues are finally biotinylated. Afterwards, the biotinylated proteins can be detected with anti-biotin antibodies or can be purified by affinity chromatography on neutravidin agarose. In this chapter, we give a detailed description of the biotin switch method, which can be used for proteomics approach to identify candidates for protein S-nitrosylation as well as to analyse S-nitrosylation of selected proteins.

**Key words:** Biotin switch, S-nitrosylation, nitric oxide.

## 1. Introduction

Nitric oxide (NO) is a free-radical product of cellular metabolism in microorganisms, plants, and animals that is involved in many different physiological processes, such as defense, growth and development, neurotransmission, vasodilation, and inflammation *(1–9)*. Many of the biological functions of NO arise as a direct consequence of chemical reactions between proteins and NO or NO oxides generated as $NO/O_2$ or NO/superoxide reaction products. The reactions of NO with metal ions of heme groups or the forma-

John T. Hancock (ed.), *Methods in Molecular Biology, Redox-Mediated Signal Transduction, vol. 476*
© 2008 Humana Press, a part of Springer Science + Business Media, Totowa, NJ
DOI: 10.1007/978-1-59745-129-1_15

tion of dinitrosyl complexes are demonstrated to play important roles in NO signaling. The best known example is soluble guanylate cyclase. This heme protein is activated by NO and catalyses the formation of cyclic GMP, which acts as second messenger in various biological responses *(10)*. However, during the last decade, protein S-nitrosylation—the covalent attachment of NO to sulfhydryl groups of cysteine residues—moved to the center of importance of NO-dependent signaling *(11–14)*. This type of protein modification is a reversible redox-based enzyme-independent mechanism, which is involved in regulation of many physiological processes. S-nitrosylation of cysteine residues can alter the activity of enzymes and transcription factors, can result in translocation of the modified proteins, or can change its physiological function *(15–18)*. However, because of the lability of nitrosothiols, it was difficult to detect and analyze this type of protein modification. Jeffrey and colleagues developed a method, named the biotin switch method, which overcome this limitation *(19, 20)*. The principle behind this method is the substitution of the NO group for a biotin linker (**Fig.15.1**). In this way, the labile S–NO bond is replaced

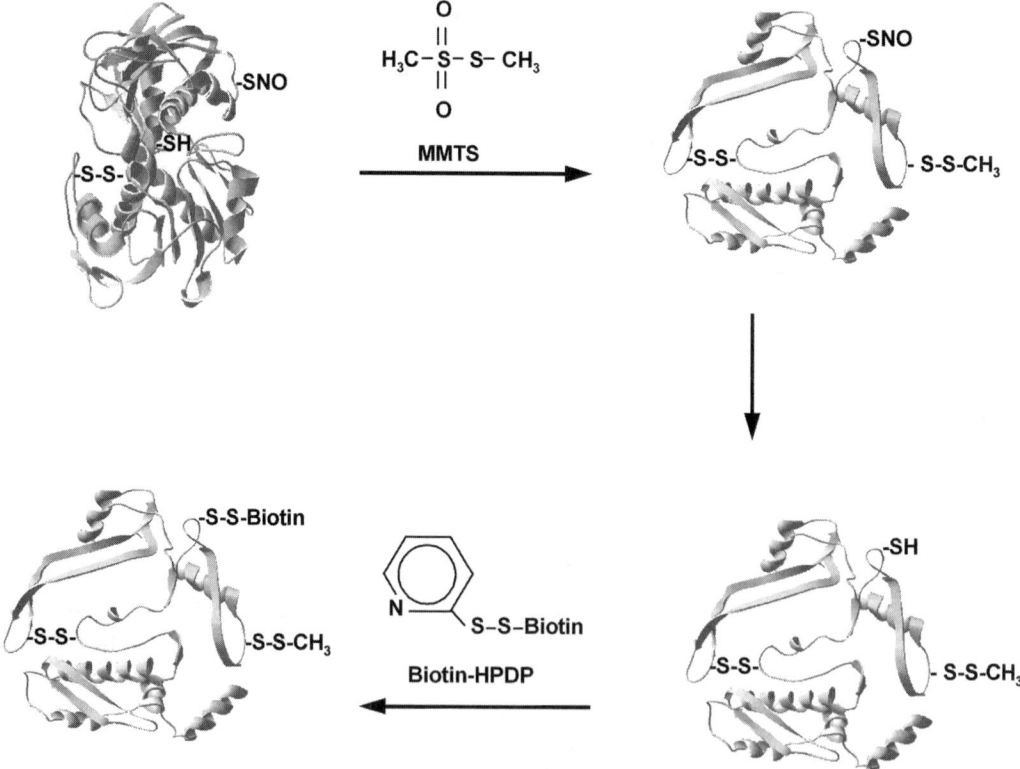

Fig. 15.1. Schematic presentation of the biotin switch assay. A model protein containing cysteine residues in the disulfide, free thiol, and nitosothiol stage is subjected to the biotin switch assy. First, the free thiol group is blocked with MMTS in the presence of SDS, which ensures the access of buried thiol groups. Then nitrosylated thiols are reduced selectively by ascorbate. In the third step, the re-formed thiol groups are biotinylated with the thiol-modifying agent biotin-HPDP.

by a more stable disulfide bridge. This biotin-labeling allows the detection and purification of the previously S-nitrosylated proteins using anti-biotin antibodies or immobilised streptavidin, respectively. The biotin switch method was successfully used to identify candidates for S-nitrosylation in different organisms and tissues, such as rat brain lysate, mouse mesangial cells, rat liver mitochondria, endothelia cells, *Mycobacterium tuberculosis* and *Arabidopsis thaliana* leaves and cell cultures *(19, 21–25)* (**Figs. 15.2, 15.3 and 15.4**). However, there are also some reports doubting the specificity of the biotin switch method *(26–28)*.

## 2. Materials

### 2.1. Generation of S-Nitrosothiols

1. HEN buffer for protein extraction: 25 mM N-(2-hydroxyethyl)piperazine-N'-2-ethanesulfonic acid (HEPES)-NaOH, pH 7.7; 1 mM ethylene diamine tetraacetic acid (EDTA); and 0.1 mM neocuproine (Sigma). Buffers containing neocuproine should be protected from light.

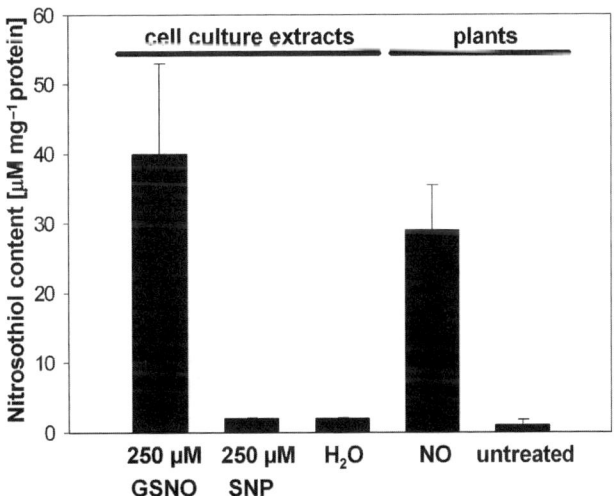

Fig. 15.2. Nitrosothiol content in *Arabidopsis thaliana* cell suspension culture extracts after treatment with NO donors and in leaves of NO-treated *A. thaliana* plants. After treatment of cell culture extracts with either 250 µM of GSNO, 250 µM of SNP, or water, remaining NO donors were removed by the use of chromatography on a Sephadex G-25 M column. Nitrosothiol contents were determined according to Saville *(29)*. Additionally, nitrosothiol content from leaves of *A. thaliana* plants treated with NO gas and from leaves of untreated plants were measured. Treatment of cell culture extracts with 250 µM of GSNO resulted in a nitrosothiol content of approx 40 µM/mg protein. The nitrosothiol content in *Arabidopsis* leaves treated with NO gas reached values up to 30 µM/mg protein. Values represent mean of at least two independent determinations.

2. *S*-Nitrosoglutathione (GSNO) is dissolved in water at 10 m*M* and stored in aliquots at –20°C. GSNO decomposes rapidly in solution (approx. 5% per hour in water at room temperature). To avoid faster decomposition the dissolved GSNO should be protected from light and from contact with transition metals, especially Cu(I).

3. PD10 columns from Amersham Bioscience, (Freiburg) and Micro Bio-spin P6 columns from Biorad to get rid of low molecular weight compounds (<6,000 kDa).

### 2.2. Biotin-Switch

1. 2 *M* Methyl methanethiosulfonate (MMTS) in dimethylformamide.

2. 50 m*M* ascorbate in water; stored in aliquots at –20°C.

3. *N*-[6-(biotinamido)hexyl]-3′-(2′-pyridyldithion)propion amide (Biotin-HPDP) from Perbio (Bonn) dissolved in dimethylformamide at a concentration of 4 m*M* and stored at –20°C.

4. HENS buffer: 25 m*M* HEPES-NaOH, pH 7.7; 1 m*M* EDTA; 0.1 m*M* neocuproine; and 1% sodium dodecyl sulfate (SDS).

### 2.3. SDS-Polyacrylamide Gel Electrophoresis

1. Separation buffer (4X): 1.5 *M* Tris-HCl, pH 8.7; 0.4% SDS.

2. Stacking buffer (4X): 0.5 *M* Tris-HCl, pH 6.8; 0.4% SDS.

3. Ten percent ammonium persulfate (APS) in water. Store aliquots at –20°C or use always freshly prepared solutions of ammonium persulfate.

4. *N,N,N,N′*-Tetramethyl-ethylenediamine (TEMED) (Serva).

5. Prestained molecular weight markers (Fermentas).

6. Thirty percent polyacrylamide solution containing 0,8% bisacrylamide.

7. SDS-PAGE sample buffer: 50 m*M* Tris-HCl, pH 6.8; 2% (w/v) SDS; 20% (v/v) glycerol; and 0.01% (w/v) bromphenol blue. Store at room temperature. If proteins should be separated under reduced conditions add 2 mM dithiothreitol (DTT) just before use.

8. Ten percent (w/v) SDS.

9. Running buffer: 25 m*M* Tris, 0.2 *M* glycine, 0.1% (w/v) SDS.

### 2.4. Western Blotting

1. Towbin Transfer buffer: 25 m*M* Tris, 192 m*M* glycine, 0.1% (w/v) SDS, 20% (v/v) methanol. Store transfer buffer at 4°C and add methanol just before use.

2. PVDF-membrane (0.2 μm) was obtained from Pall (Portsmouth, UK).

3. Nitrocellulose membrane (0.2 μm) was purchased from Schleicher & Schuell (Dassel, Germany).

4. Tris-buffered saline (TBS): 10X stock solution—100 m*M* Tris-HCl, pH 7.4; 1.5 *M* NaCl; and 10 m*M* MgCl$_2$. Dilute 100 mL with 900 mL of water for use.

5. Tris-buffered saline with Tween (TBS-T): 0.5% (v/v) Tween 20 in TBS.

6. Blocking buffer: TBS-T supplemented with 1% (w/v) non-fat dry milk and 1% (w/v) BSA.

7. Mouse anti-biotin antibody coupled with alkaline phosphatase (Invitrogen).

8. A solution of 10% nitroblue tetrazolium chloride (NBT) (Roth) is prepared in 70% dimethylformamide.

9. A solution of 5% 5-bromo-4-chloro-3-indolyl phosphate (BCIP) (Roth) is prepared in water.

10. Gel Blotting Paper GB002 from Whatman (Schleicher & Schuell).

**2.5. Purification of Biotinylated Proteins**

1. Empty columns (BioRad, Munich).

2. Immobilized Neutravidin (Perbio, Bonn).

3. Neutralisation buffer: 20 m*M* HEPES-NaOH, pH 7.7; 100 m*M* NaCl; 1 m*M* EDTA; and 0.5% (v/v) Triton X-100.

4. Washing buffer: 20 m*M* HEPES-NaOH, pH 7.7; 600 m*M* NaCl, 1 m*M* EDTA; 0.5% (v/v) Triton X-100.

5. Elution buffer: 20 m*M* HEPES-NaOH, pH 7.7; 100 m*M* NaCl, 1 m*M* EDTA, 100 m*M* β-mercoptoethanol (add β-mercoptoethanol just before use).

# 3. Methods

To identify candidates for S-nitrosylation, it is recommended that one use freshly prepared protein extracts of the organism or tissue of interest. Moreover, the method is also suitable to analyse S-nitrosylation of purified proteins. It is important that proteins are in their native state; otherwise, unspecific S-nitrosylation may occur. Proteins should be extracted in HEN buffer, which contains EDTA and neocuproine. Especially, the addition of neocuproine is important to protect the nitrosothiols from copper-dependent degradation. Furthermore, the assigned timeframe shall be kept for the different steps of the procedure, because of the lability of the nitrosothiols. If other buffer systems have to be used for protein extraction, buffers can be exchanged using gel filtration columns (e.g., P-6 micro bio-spin columns or PD-10 desalting columns). If reducing agents such as DTT, β-mercaptoethanol, or glutathione (GSH) are necessary for protein extraction, ensure that these low molecular weight thiols have been removed before S-nitrosylation. For immunoblot analyses start with about 100 µg protein, whereas 10–50 mg of protein should be used, if identification of S-nitrosylated proteins is attended. This will result in approximately 50–250 µg of purified protein.

### 3.1. In Vitro S-Nitrosylation of Proteins

1. Isolate/prepare proteins, which are supposed to be analysed for S-nitrosylation in HEN-buffer (see **Note 1**).

2. Quantify the protein concentration, e.g., according to Bradford and adjust the protein concentration to 0.8 µg/µL HEN. This concentration guarantees an effective S-nitrosylation and subsequent blocking of free thiols, respectively.

3. Add 10–500 µ$M$ GSNO as NO donor to the protein sample and incubate at 25°C in darkness for 20–30 min. Treatment of a protein sample with GSH can be used as negative control. To avoid light-dependent decomposition of nitrosothiols the samples must be protected from light from now on (see **Notes 2** and **3**).

4. Optional: To avoid unspecific S-nitrosylation during the blocking procedure remove excessive GSNO by a gel filtration step. However, because the blocking reagent is used in

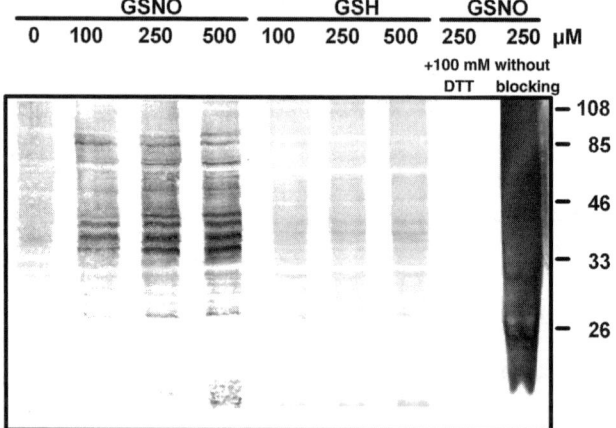

Fig. 15.3. Detection of S-nitrosylated proteins of *Arabidopsis thaliana* cell culture extracts. Extracts containing 100 µg of protein were treated with different concentrations of GSNO or GSH and labeled with biotin using the biotin switch method. Additionally, proteins were S-nitrosylated with 250 µM of GSNO and reduced with 100 m$M$ of DTT after biotinylation. The sample on the right side was treated with 250 µM of GSNO and underwent biotin switch method without MMTS treatment (blocking step). Proteins were separated by SDS-PAGE and blotted onto PVDF-membrane. Detection of biotinylated proteins was achieved using anti-biotin antibody. The relative masses of protein standards are shown on the right.

high excess, bared thiol groups almost exclusively react with the blocking reagent so that S-nitrosylation reactions during this step can be neglected. We could not detect a difference between samples with or without removing GSNO.

### 3.2. Biotinylation of Nitrosothiols

1. To block non-nitrosylated free thiols, denature the proteins and methylthiolate the thiols by adding 2% SDS and 20 mM MMTS (final concentrations) to the sample. Incubate the reaction at 50°C for 20 min and vortex every 4 min. Because nitrosothiols are temperature labile, as much as 25% nitrosothiols might be lost during the blocking step. For this reason, if you need to increase the blocking efficiency, it is recommended to increase the MMTS concentration instead of extending the incubation time.

2. The proteins are precipitated by addition of two volumes of ice-cold acetone at −20°C for at least 20 min to remove residual MMTS. Centrifuge the proteins for 10 min at 10,000*g* and 4°C and rinse the pellet and the walls of the tubes with acetone.

3. Upon drying the pellet for 5–10 min, the proteins are resuspended in HENS so that the resulting protein concentration is 10 mg/ml (corresponding to the starting material). Avoid overdrying of the protein pellet; otherwise, the resuspension might be difficult!

4. By adding 1 mM ascorbate and 1 mM biotin-HPDP, nitrosothiols are selectively reduced and the nitrosoresidues are

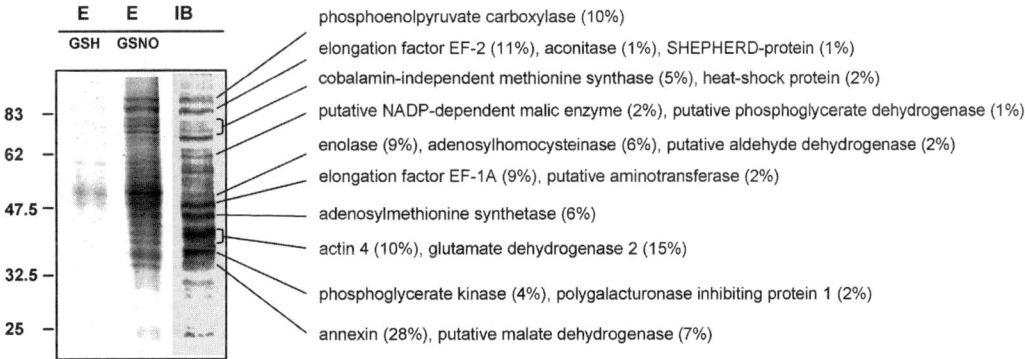

Fig. 15.4. S-nitrosylated proteins of *Arabidopsis thaliana* cell cultures. 10 mg of cell culture proteins were treated with 250 μM of GSH or GSNO and subjected to the biotin switch method. Biotinylated proteins were purified with affinity-chromatography by using neutravidin-agarose. Eluates (E) were separated with SDS-PAGE and visualized with the use of Coomassie staining. Protein bands corresponding to predominant bands of the immunoblot analysis (IB) were identified with nanoLC/MS/MS. The percentage of protein covered by the matched peptides is given in brackets. The relative masses of protein standards are shown on the *left*.

exchanged with biotin during a 1 h incubation at 25°C (it is not longer necessary to protect the samples from light). Further controls might be useful: Without ascorbate or biotin-HPDP respectively or with DTT treatment after biotinylation (*see* **Notes 4** and **5**).

5. To detect the biotinylated proteins separate them under non-reducing conditions (*see* Sect. 3.4) and transfer the proteins onto PVDF or nitrocellulose membrane (*see* Sect. 3.5). Biotinylated proteins can be detected using anti-biotin antibodies (*see* Sect. 3.6.). For purification of biotinylated proteins, precipitate the proteins with two volumes of acetone at –20°C to remove excess biotin-HPDP. Recover the proteins by centrifugation, wash the pellet with acetone, resuspend it in HENS buffer (~10 mg/mL), and follow the instructions in Section 3.3.

### 3.3. Neutravidin Purification of Biotinylated Proteins

1. Equilibrate 30 µL of the 50% Neutravidin-agarose slurry per milligram of protein with neutralization buffer.

2. Add at least two volumes of neutralization buffer to the biotinylated proteins in HENS buffer as well as the equilibrated Neutravidin-agarose. Because the use of SDS can decrease the Biotin-Neutravidin binding efficiency, up to 20 volumes of neutralization buffer can be added to the biotinylated proteins to dilute the SDS-concentration of the HENS buffer. Incubate at 25°C for 1 h with gently shaking.

3. Centrifuge the samples for 1 min at 200$g$, discard the supernatant, wash the matrix two times with ten matrix volumes of washing buffer. Centrifuge for 1 min at 200$g$ between each washing step. Transfer the matrix into an empty column and perform a third washing step on the column.

4. Close the column, add at least two matrix volumes of elution buffer and eluate the bound proteins after incubation for 5 min. β-mercaptoethanol in the elution buffer reduces the disulfide bridge between the biotin and the purified protein resulting in the release of the protein from the biotin linker.

5. Precipitate the purified proteins with two volumes of acetone at –20°C preferably over night, recover the protein by centrifugation (30 min at 4°C and 25000x$g$) and follow up the reducing SDS-PAGE.

### 3.4. SDS-page

1. The following instructions are based on the Hoeffer SE-250 mini-vertical unit (8 × 10 cm). Glass and notched alumina plates are cleaned with double distilled water and finally with 70% ethanol.

2. Assemble the glass and alumina plates with 1-mm spacers in-between and set them into the gel caster.

3. Prepare a 1.0-mm thick, 10% separation gel by mixing 1.65 mL of 30% polyacrylamide containing 0.8% bisacrylamide, 2.02 mL of ddH$_2$O, 1.25 mL of Separation buffer (4X), 50 µL of 10% SDS, 2.5 µL of TEMED, and 25 µL of 10% APS. TEMED is used together with APS to catalyze the polymerization of acrylamide. Therefore, it is important to add both components directly before preparing the gel.

4. Pour the separating gel, leaving about 3 cm space for the stacking gel and overlay with isopropanol. After polymerization (30–60 min), exchange isopropanol with double-distilled water. The separation gel can be stored overnight at 4°C.

5. Directly before pouring the stacking gel discard the water and remove residual liquid with Whatman paper. Take care not to touch the gel surface.

6. Insert an appropriate comb and prepare the stacking gel by mixing 0.5 mL of 30% polyacrylamide solution, 1.25 mL of ddH$_2$O, 625 µL of stacking buffer (4X), 25 µL of 10% SDS, 10 µL of TEMED, and 20 µL of 10% APS. Immediately pour the stacking gel, because polymerization occurs quickly (ca. 5 min).

7. Fix the gel with clamps in the running chamber, remove the comb, fill the upper and the lower buffer reservoir with SDS running buffer, and rinse the wells with buffer to remove nonpolymerized acrylamide.

8. Resuspend protein pellets in the appropriate SDS-PAGE sample buffer. For separation of biotinylated proteins, it is important to omit reducing agents, such as DTT or β-mercaptoethanol, in the sample buffer and do not boil the samples, otherwise the biotin linker will be lost. In contrast, the Neutravidin purified proteins are dissolved in reducing sample buffer and boiled at 95°C for 5 min before loading on the gel.

9. Apply the samples as well as 5 µL of prestained protein marker on the gel, close the lid, connect the electrophoresis unit to a current generator and apply 25 mA per gel. The separation of the proteins takes about 60–90 min.

10. Switch off the power supply, disconnect the electrophoresis unit, and disassemble the glass/alumina plates. Cut off the stacking gel and one corner of the separating gel for later assignment of the lanes on the gel/blot and the accordant samples.

11. Depending on the protein amount, different protein staining techniques are available. The detection limit of Coomassie blue is in the range of 50 ng protein, whereas staining

with silver or Sypro ruby allows detection of less than 1 ng protein. To transfer the biotinylated proteins from the SDS gel onto a PVDF or nitrocellulose membrane proceed with Sect. 3.5.

## 3.5. Western Blotting of Biotinylated Proteins

1. The transfer of the separated proteins is carried out with the use of a Graphic Electroblotter.

2. During the protein separation, prepare nine sheets of 3-mm paper as well as one piece of membrane (PVDF or nitrocellulose) in the size of the separating gel

3. PVDF membranes have to be activated before use. The activation is done in methanol for 5 s, followed by a washing step in ddH$_2$O for 5 min and the final equilibration of the membrane in transfer buffer for at least 15 min (the nitrocellulose membrane should also be equilibrated). Take care that the membranes once activated and equilibrated do not get dry! Nitrocellulose membranes just have to be incubated in transfer buffer for 15 min (*see* **Note 6**)

4. Let the Whatman paper sheets soak with transfer buffer

5. Set up the blot from anode to cathode: Six sheets of Whatman paper, the equilibrated membrane, the gel upside up, on the top three sheets of Whatman paper, and finally the lid with the cathode. Ensure that no bubbles are trapped in the sandwich

6. Connect the electroblotter with the power supply and apply the calculated current for 1 h. Calculation of the current: gel area multiplied by 2.5 (length in cm × width in cm × 2.5)

7. After the transfer is complete disconnect the electroblotter from the power supply, carefully lift off the lid, remove the upper three 3-mm paper sheets, and cut off the same corner of the membrane that is cut off from the gel for orientation. The prestained standard proteins can be used as transfer control and should be visible on the membrane

8. The membrane should be directly followed up as described in Sect. 3.6.

## 3.6. Immunodetection of Biotinylated Proteins

1. Transfer the membrane into a tray with 30 mL of blocking buffer. The blocking lasts 1 h at 25°C or over night at 4°C. Blocking and the following steps should be conducted by gently shaking.

2. The blocked membrane is incubated in 10 ml TBS-T + 1% (w/v) nonfat dry milk with 1 µL of the anti-biotin antibody conjugated with alkaline phosphatase (final dilution: 1:10,000) for 1 h at 25°C.

3. Remove the antibody solution (*see* **Note** 7) and wash the membrane two times with 20 mL of TBS-T and once with TBS each for 10 min at 25°C.

4. Prepare the developing solution directly before use: 10 mL of AP buffer supplemented with 33 µL of NBT and 33 µL of BCIP.

5. Transfer the washed membrane on a clean glass plate and pour the developing solution on the membrane (without rocking).

6. Wait until the detected signals are strong enough but for a maximum of 10 min. The longer the development, the stronger the background will arise.

7. To stop the development, rinse the membrane with ddH$_2$O.

---

## 4. Notes

1. For *in vitro* S-nitrosylation experiments, it is important that the isolated proteins are in their native state; otherwise, unspecific S-nitrosylation may occur.

2. Because GSNO is a physiological NO donor/reservoir, it is appropriate to *in vitro* S-nitrosylation of proteins. Alternatively, 1-[2-(2-aminoethyl)-*N*-(2-ammonioethyl)amino] diazen-1-ium-1,2-diolate (DETA NONOate), *S*-nitroso-*N*-acetylpenicillamine (SNAP), or sodium nitroprusside (SNP) often are used as NO donors.

3. Furthermore, some laboratories might have the opportunity/equipment to expose their biological samples to gaseous nitric oxide. For analyses of *in vivo* S-nitrosylation, effective endogenous NO production have to be guaranteed.

4. Nitrosothiols are exceptionally labile. Especially in the presents of light and reduced metal ions, such as Cu$^+$, the half-lives of nitrosothiols is in the range of seconds to a few minutes. Therefore metal-complexing compounds like EDTA and neocuproine are essential buffer additives to protect nitrosothiols from degradation.

5. The use of ascorbate for reduction of S-nitrosothiols is analyzed critically. It was demonstrated that greater concentrations and longer incubation times can significantly improve immunological detection of S-nitrosylated proteins resulting in false-positive signals. Therefore, particular attention should be kept on the ascorbate-dependent reduction step of the biotin switch to avoid false-positive signals.

6. In samples that were not treated with ascorbate and/or biotin-HPDP as well as in samples that were reduced with DTT after the biotin modification, the endogenously biotinylated proteins are detectable.

7. Transfer buffer: For protein transfer commonly Towbin buffer containing 5–20% methanol is used. The use of greater methanol concentrations decrease the transfer efficiency but increase the binding of proteins to the membrane and prevent gel swelling with heating. Greater SDS concentrations decrease the binding of proteins to the membrane but facilitate extraction of proteins from the gel.

8. For economy, the antibody solution can be reused up to three times. It can be stored at 4°C up to 2 wk; for long-term storage, it should be kept at –20°C.

## References

1. Tuteja, N., Chandra, M., Tuteja, R., and Misra, M. K. (2004) Nitric oxide as a unique bioactive signaling messenger in physiology and pathophysiology. *J Biomed Biotechnol* **2004**, 227–237.

2. Wendehenne, D., Durner, J., and Klessig, D. F. (2004) Nitric oxide: a new player in plant signaling and defence responses. *Curr Opin Plant Biol* **7**, 449–455.

3. Shapiro, A. (2005) Nitric oxide signalling in plants. *Vitam Horm* **72**, 339–398.

4. Bove, P. F., and van der Vliet, A. (2006) Nitric oxide and reactive nitrogen species in airway epithelial signaling and inflammation. *Free Radic Biol Med* **41**, 515–527.

5. Cohen, R. A., and Adachi, T. (2006) Nitric-oxide-induced vasodilatation: regulation by physiologic S-glutathiolation and pathologic oxidation of the sarcoplasmic endoplasmic reticulum calcium ATPase. *Trends Cardiovasc Med* **16**, 109–114.

6. Grun, S., Lindermayr, C., Sell, S., and Durner, J. (2006) Nitric oxide and gene regulation in plants. *J Exp Bot* **57**, 507–516.

7. Mannick, J. B. (2006) Immunoregulatory and antimicrobial effects of nitrogen oxides. *Proc Am Thorac Soc* **3**, 161–165.

8. Moncada. S., Bolanos, J. P. (2006) Nitric oxide, cell bioenergetics and neurodegeneration. *J Neurochem* **97**, 1676–1689.

9. Villalobo, A. (2006) Nitric oxide and cell proliferation. *FEBS J* **273**, 2329–2344.

10. Russwurm, M., and Koesling, D. (2004) NO activation of guanylyl cyclase. *EMBO J* **23**, 4443–4450.

11. Lane, P., Hao, G., and Gross, S. S. (2001) S-nitrosylation is emerging as a specific and fundamental posttranslational protein modification: head-to-head comparison with O-phosphorylation. *Sci. STKE* **June 12**(86), RE1.

12. Stamler, J. S., Lamas, S., and Fang, F. C. (2001) Nitrosylation. the prototypic redox-based signaling mechanism. *Cell* **106**, 675–683.

13. Gaston, B. M., Carver, J., Doctor, A., and Palmer, L. A. (2003) S-nitrosylation signaling in cell biology. *Mol Interv* **3**, 253–263.

14. Hess, D. T., Matsumoto, A., Kim, S. O., Marshall, H. E., and Stamler, J. S. (2005) Protein S-nitrosylation: purview and parameters. *Nat Rev Mo. Cell Biol* **6**, 150–166.

15. Hausladen, A., Privalle, C. T., Keng, T., DeAngelo, J., and Stamler, J. S. (1996) Nitrosative stress: activaton of the transcription factor OxyR. *Cell* **86**, 719–729.

16. Marshall, H. E., and Stamler, J. S. (2001) Inhibition of NF-kappa B by S-nitrosylation. *Biochemistry* **40**, 1688–1693.

17. Hara, M. R., Agrawal, N., Kim, S. F., Cascio, M. B., Fujimuro, M., Ozeki, Y., Takahashi, M., Cheah, J. H., Tankou, S. K., Hester, L. D., Ferris, C. D., Hayward, S. D., Snyder, S. H., and Sawa, A. (2005) S-nitrosylated GAPDH initiates apoptotic cell death by nuclear translocation following Siah1 binding. *Nat Cell Biol* **7**, 665–674.

18. Lindermayr, C., Saalbach, G., Bahnweg, G., and Durner, J. (2006) Differential inhibition of Arabidopsis methionine adenosyltransferases by protein s-nitrosylation. *J Biol Chem* **281**, 4285–4291.

19. Jaffrey, S. R., Erdjument-Bromage, H., Ferris, C. D., Tempst, P., and Snyder, S. H. (2001) Protein S-nitrosylation: a physiological signal for neuronal nitric oxide. *Nat Cell Biol* **3**, 193–197.

20. Jaffrey, S. R., and Snyder, S. H (2001) The biotin switch method for the detection of S-nitrosylated proteins. *Sci. STKE* **June 12**(86), PL1.

21. Kuncewicz, T., Sheta, E. A., Goldknopf, I. L., and Kone, B. C. (2003) Proteomic analysis of s-nitrosylated proteins in mesangial cells. *Mol Cell Proteomics* **2**, 156–163.

22. Foster, M. W., and Stamler, J. S. (2004) New insights into protein S-nitrosylation: Mitochondria as a model system. *J Biol Chem* **279**, 25891–25897.

23. Martinez-Ruiz, A., and Lamas, S. (2004) Detection and proteomic identification of S-nitrosylated proteins in endothelial cells. *Arch Biochem Biophys* **423**, 192–199.

24. Lindermayr, C., Saalbach, G., and Durner, J. (2005) Proteomic identification of s-nitrosylated proteins in Arabidopsis. *Plant Physiol* **137**, 921–930.

25. Rhee, K. Y., Erdjument-Bromage, H., Tempst, P., and Nathan, C. F. (2005) S-nitroso proteome of *Mycobacterium tuberculosis*: Enzymes of intermediary metabolism and antioxidant defense. *Proc Natl Acad Sci USA* **102**, 467–472.

26. Zhang, Y., Keszler, A., Broniowska, K. A., and Hogg, N. (2005) Characterization and application of the biotin-switch assay for the identification of S-nitrosated proteins. *Free Radic Biol Med* **38**, 874–881.

27. Huang, B., and Chen, C. (2006) An ascorbate-dependent artifact that interferes with the interpretation of the biotin switch assay. *Free Radic Biol Med* **41**, 562–567.

28. Shi, Q., Chen, H. F., and Lou, Y. J. (2006) Further evidence that rat liver microsomal glutathione transferase 1 is not a cellular protein target for S-nitrosylation. *Chem Biol Interact* **162**, 228–236.

29. Saville, B. (1958) A scheme for the colorimetric determination of microgram amounts of thiols. *Analyst* **83**, 670–672.

# INDEX